STOCKMANSHIP

Improving the care of the pig and other livestock

The Forgotten Pillar...

STOCKMANSHIP

Improving the care of the pig and other livestock

Peter English
Gethyn Burgess
Ricardo Segundo
John Dunne

FARMING PRESS

First published 1992

ISBN 0 85236 236 6

A catalogue record for this book is available from the British Library

Published by Farming Press Books, Wharfedale Road, Ipswich IP1 4LG, United Kingdom

Distributed in North America by Diamond Farm Enterprises, Box 537, Alexandria Bay, NY 13607, USA

Cover design by Andrew Thistlethwaite
Typeset by Galleon Photosetting
Printed in Great Britain by Jolly & Barber Ltd

Contents

**A colour section appears between
pages 114 and 115**

Acknowledgements

The interest in and enthusiasm for the subject of this book emanates from long and varied contacts with hosts of stockpeople caring for all the farm animal species. We have gleaned something from each and all of these animal carers, from the most basic of small-holdings to those on the most sophisticated large farms, in all parts of the world. We acknowledge this debt.

We also owe much to students over the years, especially for delving, in their enthusiasm for new knowledge, into such aspects as the contribution of stockmanship in animal production systems, the components of stockmanship, the factors influencing job satisfaction for stockpeople, approaches to training in stockmanship and methods of evaluating the effectiveness of training. The enthusiasm of these youngsters for the subject and their appreciation of the great skills and qualities of good stockpeople have, through lively discussion and research, helped to keep our own minds focused on this topic and have contributed significantly to the production of this book.

While we and our students at the University of Aberdeen were engaged in our studies into stockmanship, fellow researchers elsewhere in the world were making major contributions. Notable among these was Dr Martin Seabrook, now at the University of Nottingham, who in his early years at the University of Reading worked with the late Mr Rex Patterson in investigating the latter's many one-man dairy units which were almost identical apart from the operating stockmen. Martin Seabrook grasped this great opportunity to estimate the impact of the stockpeople on milk production on these individual farms and proceeded to evaluate the characteristics and attitudes of high and low achieving cowmen. Martin has shown great determination over the years to progress his studies on many aspects of the art and science of stockmanship. He and his co-workers have achieved a great deal despite the totally inadequate funding available for this most important component of animal production systems.

Another significant contributor was Dr Paul Hemsworth, who appeared on the 'stockmanship' scene in the late 1970s. He and his colleagues in Holland, the USA and especially in Australia, have furthered our understanding of the subject greatly in the last decade. More recently Paul and his colleagues have begun to apply the results of their researches on Australian pig farms to the benefit of the stockpeople, the welfare, health and productivity of the pigs and the efficiency of the businesses involved. The Australian Pig Science Association deserves great credit for their encouragement and financial support for these notable research and development studies.

We recognise the great contribution to the training of stockpeople by lecturing and farm staff involved in training of prospective stockpeople at agricultural colleges and the commendable efforts of organisations such as the Agricultural Training Board in Britain. Full-time, part-time and day release classes for stockpeople have contributed greatly to maintaining the quality of stockmanship and to the job satisfaction of the recipients.

Nearer home, we would like to acknowledge our debt to the many farmers, managers and stockpeople in the North of Scotland and elsewhere who have worked with us in our researches and in applying acquired knowledge and increased understanding of stockmanship at farm level. We have the greatest of admiration for the stockpeople themselves—animal lovers almost without exception and with great aspirations for improved understanding and enhanced husbandry of the livestock in their care. These honest and honourable workers are 'the salt of the earth'.

We are most grateful for the good 'stockmanship' accorded to us by colleagues who fuel and stimulate both our interests in and our efforts to improve the qualities and job satisfaction of stockpeople— Owen MacPherson being the top candidate for our 'Stockman of the Era' award. Owen was also responsible for many of the photographs. We would also like to thank Donald Macdonald, Sandra Edwards, Mike Birnie, Margaret Warren, Grant Myles and a host of other colleagues for their contributions and support.

To those who provided photographs and helped in many other ways we are most indebted. These included Alf, Ken, Malcolm and Donald Rae, Arthur Simmers, Forbes Davidson, Andy Brown, Anne MacKenzie, George Marr, Alfie Marr, Cecil Morris, George Bruce, the Cranfield 'team' of John, Lorna, Gillian and Graham Smith, Howard and Catherine Davies, Marysia Stamm, Janet Roden, Jim Marr, Euan

Hart, Ernest Auton, Bert Kelman, Jimmy MacLeod, Henry and Phil Dunne, May Barbour, Helen Burgess, Rhona Law, Donald Mackay, the Strontoiller duo of Flora and Malcolm MacGregor, Peter Menzies, Gonzalo Castro, Tarsicio Brebiesca, Pedro Brebiesca, Joaquin Becerril, Luis Fernando Ruiz, Francisco Villegas, Dorothy Kidd at the National Museum of Scotland and the staff at Aberdeen Art Gallery. The staff of the Aberdeen University Graphics Department were very supportive and we much appreciate the skills and efforts of Martin Cooper and his colleagues. We are also grateful to all those organisations and individuals who allowed us to reproduce their photographs.

Mrs Joey Parker has prepared the drafts in her usual highly efficient manner. Her natural, joyful charm and presence have never failed to uplift us even when the going was toughest. Many thanks, Joey, for keeping on smiling even when we were in our most impatient and irritable moods.

Finally, another lady in the same mould is Julanne Arnold at Farming Press. Julanne and her colleagues worked wonders with our jumble of drafts, tables, figures and photographs in evolving the final product of which we are proud and which we hope will be a useful contribution to a field which deserves much more attention and support than it has received to date from official bodies and people in positions of influence.

PRE, GB, RSC, JHD
Aberdeen
April 1992

To caring stockpeople everywhere—the world's most undervalued profession.

Also to the authors' domestic carers—whose patience and devotion in the face of much neglect and provocation displayed all the hallmarks of excellent stockmanship.

Foreword

In modern animal production the role and importance of the stockperson has been neglected. While it is appreciated in some sections of the animal industry that the technical skills and knowledge of stockpeople have a large impact on animal efficiency, health and welfare, the role of other human factors such as the attitude and behaviour of the stockperson towards farm animals and the work ethic of the stockperson, has in general been ignored by the animal industries, animal scientists and agricultural educators. This is perhaps not surprising with the attainments and benefits which have been achieved in modern animal production as a consequence of rapid improvements in knowledge and technology in areas such as nutrition, genetics, health and environmental control. The basic, practical and less glamorous subject of stockmanship has been overlooked or ignored in this rush to achieve these rapid improvements in animal productivity obtainable from other initiatives.

Nevertheless, there has been some effort, albeit a relatively small one, by animal scientists and industry personnel to address this important imbalance. Research on the impact and components of stockmanship is therefore in its infancy, but this research to date in several animal industries has indicated that human factors may have large and at times surprising consequences on animal efficiency, health and welfare.

For further research and development to occur in this area of stockmanship, it is important for animal scientsts, key industry personnel and educators to recognise the importance of this subject. Our current understanding of this subject, although limited, should be sufficient to stimulate this recognition. This book, by reviewing recent research by students and established scientists, is an important step in creating an increased awareness of the importance of stockmanship.

This book will also provide farm managers and stockpeople with general guidance on ways in which stockmanship may be improved. There are several other benefits that will accrue as a result of developments such as this book: a recognition that stockpeople have an important pivotal role in animal efficiency, health and welfare which is likely to lead to long-term gains in work performance through improvements for stockpeople in areas such as self-esteem, working conditions, remuneration, training and career opportunities. Therefore I commend this book in the belief that initiatives of this nature will stimulate a number of developments in the area of stockmanship which are long overdue and which have important practical implications for the health, welfare and performance of our farm animals.

DR PAUL HEMSWORTH
Victorian Institute of Animal Science
Department of Food & Agriculture
Victoria
Australia

STOCKMANSHIP

Improving the care of the pig and other livestock

1

Introduction

People are the most important economic resource of a society. The main factor in the determination of a society's material well-being is the strength of its human resources—that is, its people.

This book is about a specific group of people: those working in animal production systems and particularly with the pig. Stockmanship is a vital component of all animal production systems. It affects the following:

- the welfare of the animals in the system
- the quantity and quality of the output of the system
- the productivity, efficiency and competitiveness of the system
- the adaptability and flexibility of the system

Moreover, stockpeople form a part of the rich fabric of a modern rural community and contribute to the vitality of a rural economy and rural life.

One of the purposes of this book is to accord a higher status to stockpeople than they enjoy at present and to set the great importance of stockmanship in a true perspective. Stockmanship is difficult to define in a short sentence. The typical dictionary definition of the term 'stockman' as 'a man engaged in the breeding and rearing of farm livestock' is open to criticism on the grounds of suggesting a simple and narrow role for the stockman as against the reality of a complex and multi-faceted role. The approach taken in this book is to examine, in turn, the various components of good stockmanship. It is intended that as the book proceeds the reader will gain a better understanding of the vital part played by stockmanship in animal production systems and of the aspects that contribute to excellence in stockmanship.

1

The main objectives of this book are:

- to focus on the substantial size of the stockmanship profession in both British and European agriculture
- to highlight the importance of stockmanship in livestock farming
- to emphasise the role of empathy in stockmanship
- to consider factors which contribute to excellence in the quality of stockmanship
- to explore ways of improving stockmanship through the job application and selection process
- to consider ways of improving stockmanship through the retention of good stockpeople
- to develop the basis for a culture of training in stockmanship within the farm animal industry; in other words, to point the way to ensuring a progressive acquisition and application of knowledge and skills by people working with farm animals
- to explore ideas and procedures which might improve the job satisfaction and motivation of those people who work as stockpersons

People are the most important resource in a society, including the animal industries. Here the photographer catches 'afternoon tea-time' on the pig farm.

The stockman has a direct influence on the welfare of his pigs, their productivity, the quality of the pigs produced, the efficiency of operation of the unit and the adaptability of the system. (Reproduced by kind permission of BOCM Silcock)

The book is aimed in the first place at the stockperson who has direct influence on the animal in the production system. It is the pigman who has closest contact with the animal and has the best opportunity to develop close affinity with the pigs on any farm.

The book is also aimed at the decision makers and opinion formers in the pig industry and the wider agricultural sector, including:

- the owner-manager and the employee-manager who recruit, select and employ the stockperson and in turn deploy him in the

3

operation of the animal system
- the agricultural educators who affect the content of course syllabuses and training programmes for stockpeople
- the agricultural researchers who contribute towards the development of animal production systems which stockpeople must operate in practice
- the reporters of agricultural matters whose media coverage of issues helps to form public opinion
- the agricultural students who will be the stockpersons, managers, educators, researchers and reporters of the future

The end result, it is hoped, will be to have our farm animals looked after by better equipped people who obtain more enjoyment and satisfaction from their profession, so that in turn the animals, their attendants and the owners of the animal enterprises are all likely to be beneficiaries.

TERMINOLOGY USED

The animal carer in this book is variously referred to as stockman, stockwoman and stockperson. For convenience, it is the term 'stockman' that is most frequently used but this should be taken as being interchangeable with any other description of an animal attendant.

2

The stockman and employment trends in agriculture

On all farms labour is a vital resource; good stockmen are the key in any livestock production system.

Ever since domestication of our farm animals began, the relationship between the husbandman and the animal in his care has been very close. The biblical shepherd, in seeking better conditions for his sheep, often led them rather than drove them and this practice is still current in many countries today. In subsistence agriculture in Britain and other countries, the family and their cattle shared adjacent parts of the same accommodation. The relationship between man and beast was a symbiotic one, with the cattle being dependent on man for the provision of shelter, nutrition and general care while man in turn benefited from milk, dung for fuel (or for the enhancement of land fertility) and heat production to add to the comfort of their often common dwelling in the cold of winter.

However, there have been great changes in agriculture in general and in pig production in particular in most countries in more recent times.

Evolution from subsistence agriculture to large business enterprises

Subsistence agriculture progressed first to family businesses, in which all labour was supplied by the family, and from thence to very large pig production enterprises employing many workers.

In the family-run pig production enterprise, the children were almost automatically trained in all aspects of stockmanship from the time they progressed from crawling to toddling. They thus developed their rapport with the animals from a very early age. Their awareness of animal behaviour, of animal needs, and their handling skills developed through their own experience and via the instruction and guidance of their parents. Not only did they receive a progressive and sound training in all aspects of pig husbandry but also their vested interest in the success of the business ensured a high level of motivation for, and interest in, the work they were doing.

Gradually the family business expanded to a level where family labour was insufficient to service the needs of the pig enterprise, so that

In subsistence agriculture in Britain and other countries, the family and their cattle shared adjacent parts of the same accommodation. The relationship was symbiotic, with the cattle being dependent on man for their provision of shelter, nutrition and general care while man in turn benefited from milk, dung for fuel (or to bolster land fertility), and heat production to add to the comfort of their common dwelling in the cold of winter. The photograph above shows a smallholding in Laidhay, Dunbeath, Caithness, with the house in the middle, the byre on the right and the stable on the left. (Reproduced by kind permission of National Museums of Scotland.)

Below is the plan of a similar arrangement in the Orkney Islands, showing the direct access from the living quarters of the family to the byre and stable. (Copyright of the Royal Commission on the Ancient and Historical Monuments of Scotland)

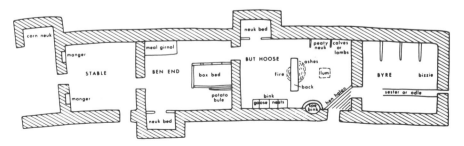

extra labour had to be hired. Some pig enterprises expanded very quickly so that many employees were now required. This created a new situation for the farmer or manager, whose upbringing had been in pig husbandry and in organising the family (self motivated and usually very biddable!) as his only labour force. Instead of being preoccupied with pig husbandry, for which he was usually eminently well trained, he now found himself acting mainly as manager of his employed personnel rather than as the husbandman of his pigs. While usually possessing plenty of common sense and ideas on how best to manage employees, perhaps largely based on general experiences in deploying family labour, he may have been ill prepared for effectively deploying a large team of paid employees.

This situation is likely to be a common one in modern agriculture. However, although the science and practices of human resource management have been applied effectively in many large industrial businesses, they have not to any great extent been applied in large scale animal production enterprises. Knowledge of human resource management must be applied more effectively to large scale pig production enterprises so that personnel can be recruited and selected more carefully, can be trained more efficiently, motivated more successfully and so that a group of such employees can be welded into a working team more effectively. Meeting these challenges forms the basis of subsequent chapters.

Traditional source of supply of stockpeople and current changes

Another trend in agriculture is worthy of note at this stage. For a long time there was a progression in agricultural employment as the farmworker's son often followed his father into the same employment. This was useful because by the time the son started his working life on the farm he had already had a long apprenticeship following at his father's coat-tails as he went about his work and benefiting, from the example shown, from the progressive instruction given and from his own personal observations and experience gained over a long period of induction. This progression into the agricultural labour force has now largely broken down because of the process of rural depopulation and the trend towards greater urbanisation. As well as the lure of the cities for entertainment purposes, the higher average wages in non-agricultural industries

7

Traditionally, the youngsters on the farm accompanied and helped the adults on all farm activities, thus gaining experience and developing skills throughout childhood so that by the time they left school and started to work on the farm they had already undergone a long apprenticeship. The picture shows bail milking in progress. (Courtesy of *Dairy Farmer*)

have contributed to this trend (see Table 2.1). Thus, a traditional and valuable source of the agricultural labour supply has been severely eroded. This means that the animal production industries have to turn to non-traditional sources for the supply of at least some of their labour. Unfortunately, guidelines on how best to attract the attention of non-traditional labour supply sources towards animal production and on how to select the best of those so attracted into the industry have not yet been set down with adequate clarity. This is another challenge which this book attempts to address.

Table 2.1 Relative weekly earnings for men (regular full time) in agriculture and non-agricultural industries in the UK (1989)

Agriculture	Non-agricultural industries	
	Manual work	Non-manual work
£167[A]	£217	£323

[A] The calculation of average wage includes the value of perquisites in the form of such items as housing and farm grown foods, e.g. milk and potatoes.

Source: Annual Abstract of Statistics, Central Statistical Office

Livestock farming in Britain—past, present and likely future position

The most notable feature of the agricultural industry in Britain is the predominance of livestock farming. At the end of the 1980s, livestock and livestock products accounted for 63% of the total value of the output of the agricultural industry (see Figure 2.1). This was despite the distorting effects of the European Community's common agricultural policy. A more detailed breakdown of the composition of the output of British agriculture shows livestock at 38% and livestock products at 25% of the total value. In the 'livestock products' category, milk dominates with 21%, followed by eggs at nearly 4% and wool at under 0.5% of the total value of agricultural output. In the 'livestock' category, the main single commodity is finished cattle and calves at 16%, followed by finished sheep and lambs at slightly more than 7%, pigs at around 7% and poultry at 6%.

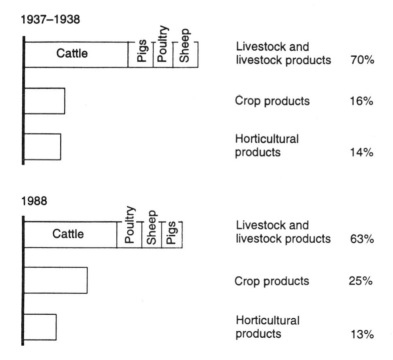

Figure 2.1   Contribution of different sectors of agriculture to the value of output of the national farm in the UK

A comparison with 50 years earlier, when Britain had a minimal agricultural policy, reveals an even greater predominance of livestock farming. In the farming year 1937–38, livestock and livestock products accounted for 70% of the total value of the output of the national farm. The breakdown then was as follows: milk 26%, cattle 16%, pigs 10%, sheep (including wool) 9%, eggs 7% and poultry 2%. Assuming that the international organisation General Agreement on Tariffs and Trade (GATT) is able in the 1990s to impose some semblance of free trade in agricultural commodities, then it is reasonable to expect some resource adjustment within British agriculture out of crop production and back into animal production. The long term prospects of Britain's agricultural industry therefore depend on efficient and competitive livestock production. This in turn will require well motivated and well trained stockpeople.

The present position in the European Community is that just over one half (54%) of the total value of agricultural output is contributed by livestock enterprises. The breakdown by commodity is milk 20%, beef 14%, pigmeat 11%, poultry 4%, eggs 4% and sheepmeat 2%. In comparison, crop products account for 20% and fruit and vegetables for 19%.

Thus, pig production is a major commodity within Britain and the EC. To service the needs of this very important sector, now and in the future, an efficient and caring workforce is essential.

Before progressing to address the important considerations of what stockmanship is, how big an impact a good stockperson can have and of how best good personnel can be selected, trained, motivated and retained in employment, some of the relevant trends over the past few decades in agriculture, as referred to earlier in this chapter, will be outlined and tabulated.

Quantification of Relevant Trends and Current Position

Trends in the size of pig enterprises in Britain over the past 4 decades are shown in Table 2.2. The trend towards fewer and larger herds is obvious, although the size of the 'national' herd has remained fairly static in the last 3 decades. Similar trends are found in many other countries.

The distribution of herd size in Britain at the time of writing is as shown in Table 2.3. Thus, while there are still many small herds in

Table 2.2   Number of pig breeding herds and average herd size in the UK

		June Agricultural Census				
		1930	1960	1970	1980	1990
Number of farms with breeding sows	(thousands)	NA*	NA	65.8	24.4	12.5
Number of breeding sows per farm		NA	NA	14	34	60
Total number of breeding sows	(thousands)	450	800	950	830	750

* NA = Data not available

Source: Agriculture in the United Kingdom (MAFF)

Table 2.3   Distribution of breeding sows by herd size in Britain (1990)

	Herd size classes (number of breeding sows per herd)		
	1–49	50–99	100 & over
Number of herds	6,883	1,184	2,154
Proportion of total breeding sows	10.4%	11.8%	77.8%

Source: Agriculture in the United Kingdom (MAFF)

Britain (less than 50 sows) these contain only a small proportion (10.4%) of the national sow herd. Around three quarters (77.8%) of sows in Britain are in herds of 100 sows and over while about 40% of the sows are in herds of 300 sows and over.

An estimate of the total number of people working with pigs in Britain is presented in Figure 2.2, and the estimated distribution of the number of employees per pig unit in Britain is summarised in Table 2.4. Of an estimated 3,725 pig herds in Britain employing one or more workers, 44.4% employ between 1 and 2 stockpeople while 55.6% employ 3 or more stockpeople.

It is clear that a large number of pig herds in Britain have a team of workers, which increases the challenge on management to weld them into an effective group as well as ensuring individual competence. Over the past few decades in Britain there has been a progressive increase in the proportion of people living in cities and large and small

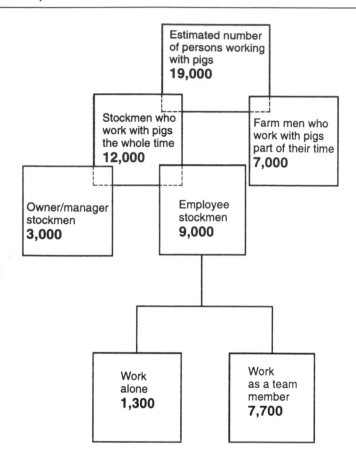

Figure 2.2   Estimated number of people working part and full time with pigs in Britain

towns relative to the proportion living in the countryside, including small villages. This trend has reduced the source of the traditional labour supply for agriculture, i.e. those from the countryside.

The changes in the proportion of the total labour force employed in agriculture relative to other industries are shown in Table 2.5.

It is clear that, over the decades, the total number of people employed in agriculture has declined, as has the proportion of the total labour force represented by agricultural workers. This situation has resulted in a decrease in another traditional source of labour supply to the agricultural industry, that is, the family members of existing agricultural employees.

Table 2.4 Distribution of pig unit employees on the basis of the size of the pig unit workforce

Size of pig unit workforce Number of paid employees(A)	Number of pig units	Number of paid employees in each size category(B)
0	7,000	
1	1,300	1,300
2	1,350	2,700
3	500	1,500
4	250	1,000
5	100	500
6	80	480
7	45	315
8	30	240
9	20	180
10	19	190
11 +	40	600
		9,005

(A) Excludes owners and part-owners
(B) Based on an estimate of 1 employee per 100 sows and progeny up to slaughter stage and one employee per 4,000 pigs in growing–finishing enterprises

Table 2.5 Trends in the labour force in the UK employed in agriculture and in all industries

		1931	1951	1980	1990
Total labour force	(thousands)	21,075	22,578	24,962	26,714
Total employed in agriculture	(thousands)	1,258	1,015	649	561
Percentage employed in agriculture	(per cent)	6.0	4.9	2.6	2.1

Source: Annual Abstract of Statistics, Central Statistical Office (1991)

Summary

The foregoing trends and current situation in Britain are very similar to those in many other countries. Thus, the supply of labour to service pig production enterprises is becoming more critical. In addition, the reduction in traditional sources of supply of pig stockpeople combined with the greater challenges of ensuring high

13

levels of stockmanship and motivation in the paid employees who are forming increasingly large teams of workers in our pig enterprises pose severe problems for pig producers.

The next chapter addresses the challenge of defining what stockmanship is and outlines its great importance in relation to animal wellbeing and the efficiency of animal production enterprises.

3

Stockmanship and its importance

The stockperson has a major influence on the health, welfare and productivity of the animals and the efficiency of the system.

Definitions of 'stockmanship'

Several people have put forward tentative definitions of stockmanship. Seabrook's (1983) definition was as follows: 'Stockmanship is knowing the behaviour pattern of animals and groups of animals within one's charge and having the ability to recognise small changes in the behaviour of any one animal or of all the animals collectively.'

English, Fowler, Baxter & Smith (1988) suggested the following more comprehensive definition: 'Good stockmanship involves a well moulded combination of:

- a sound basic knowledge of the animals and their requirements
- a basic attachment for and patience with the stock
- the ability and willingness to communicate and develop a good relationship with the stock (empathy)
- ability to recognise all individual animals and to remember their particular eccentricities
- keen sensitivity for recognising the slightest departure from normal behaviour of individual animals (perceptual skills)
- an ability to organise the working time well
- having a keen appreciation of priorities with a ready willingness to be side-tracked from routine duties as pressing needs arise to attend to individual animals in most need of attention

The basic desire is to make each animal as comfortable and contented as possible.

Attaining such objectives is also likely to be conducive to the business objective of attaining higher levels of performance and efficiency.'

15

Better definitions of 'stockmanship' will materialise when in the course of this book we increase our understanding of all that it entails.

Opinions on the importance of stockmanship

Many have expressed their opinions, largely on the basis of general observations and experience, about the importance of stockmanship in relation to animal wellbeing and animal performance.

A section of *The Report of the Technical Committee to enquire into the Welfare of Animals kept under intensive livestock husbandry systems in Britain* (Brambell, 1965) read 'we have been impressed with the importance of the standard of management and stockmanship for the welfare of farm livestock.'

Further recognition of the importance of the care of the animal by the human attendant is to be found in the *British Codes of Recommendations for the Welfare of Farm Livestock* (MAFF, 1983) as follows: 'Stockmanship is a key factor because, no matter how otherwise acceptable a system may be in principle, without competent, diligent stockmanship, the welfare of the animals cannot be adequately catered for.'

On the same theme, The British Society of Animal Production Pig Welfare Consultative Panel (BSAP, 1980) stated that:

'Good stockmanship was felt by most Panel members to have a much more important influence on animal welfare than the actual system of production, in that good stockmanship imposed on a system which was considered marginal from a welfare standpoint could lead to a high degree of comfort and contentment in the animals involved, whereas average stockmanship imposed on an acceptable system (from the welfare standpoint) could give rise to serious welfare problems.'

Recognition of the importance of the care of the animal to its performance was made by Curtis (1980) in the USA in the following terms: 'Excellent animal husbandry is the *sine qua non* of successful animal production.'

Unfortunately, although the importance of stockmanship in relation to animal welfare and animal performance appears to be widely recognised, little or no action has been taken by the relevant authorities to encourage research into stockmanship so that we can:

- increase our appreciation of all that stockmanship is
- improve our stockpeople selection procedures so as to increase the proportion of our employees in the animal production industries who possess the basic qualities of good stockmanship

- improve training methods to upgrade the quality of stockmanship
- evaluate the effectiveness of training methods as a basis for improving training procedures
- increase our understanding of how best to motivate stockpeople in order to increase levels of job satisfaction and the constancy of the care given to our farm animals
- retain high quality stockpeople more effectively on our farms

The negligible amount of research resources devoted to acquiring a better understanding of stockmanship and its improvement in Britain was highlighted by English and associates (1988) as follows:

'To date, most research and development efforts and finance in pig production have been devoted to nutrition and feeding and this has been justified on the grounds that feed constitutes about 70% of the total production costs. However, it is not an over-statement to claim that the stockman has at least a 70% influence on how efficiently or inefficiently expensive feed is converted on the pig unit into pig meat.'

The imbalance in allocation of research funds from Government in Britain continues into factors influencing animal welfare. Much finance is being invested into housing systems and components of these, with the objective of achieving improvements in farm animal welfare. While it is certainly appropriate to invest resources in this direction, the failure to sponsor research into stockmanship in relation to improving animal welfare has to be deplored.

Scientific evidence on the importance of stockmanship

1.   EVIDENCE FROM DAIRY COWS

Seabrook (1984) highlighted the influence of stockmanship in a study of 12 single-stockman dairy herds in Britain. These herds were in a single ownership, that of the late Rex Patterson, and were almost identical in terms of genotypes of cow, nutrition, buildings, facilities and general management. The most important variable differentiating them was the stockman responsible for looking after the stock and carrying out the milking in each herd. Milk production was monitored in these herds over a period of 6 years and, despite almost identical resources (apart from the stockman), there were

17

**Table 3.1   Variation in milk production of single-man dairy
herds which had almost identical cows and
management systems**

	Average annual yield per cow (litres)†						
Herd*	1965–66	1966–67	1967–68	1968–69	1969–70	1970–71	Mean
A	3232	3350	3168	3459	3064	3059	3222
B	3009	3127	3232	3364	3018	2954	3117
C	3073	3095	3304	3241	3050	2832	3099
D	3186	2841	2927	3018	2877	3013	2977
E	2945	2591	2791	2741	2791	2600	2743
F	2582	2609	2682	2809	2813	2888	2730
G	2568	2663	2991	2736	2668	2691	2720
H	2741	2786	2682	2591	2514	2600	2652
Mean of A to H	2917	2883	2972	2995	2849	2829	–
I	2373	2409	2414	2982‡	2937	2827	–
J	2995	2923	3123	3259	2163‡	2209	–
K	2595	2654	3064	3491‡	3050	3132	–
L	3082	3173	2832‡	2837	2373‡	2363	–

* Herds A to H: same cowman throughout period
  Herds I to L: change in cowman during study period
‡ Change of cowman (first complete year)
† Yield to be seen in the context of the system (low input/low output), i.e. 70 cows/herd,
  0.40 tonnes concentrate per cow per year, 0.50 ha grassland per cow (for all grazing
  and conservation), 19.8 kg nitrogen fertilizer per ha

Source: Seabrook (1984)

large between-herd differences in milk yield which obviously largely
reflected the ability of each individual stockman to care for his cows
and to milk them efficiently (see Table 3.1).

Not only was the influence of the stockman evident in comparing
the contemporary milk yields between herds (e.g. Herds A and H)
but herds experiencing a change of cowman often showed either a
marked increase (Herd I) or decrease (Herd J) in milk yield following
such a change (see Figure 3.1). Thus, in Herd I there was a change of
cowman after 3 years, and average milk yield over the next 3 years
increased to an average of 2915 litres per year (from 2399 litres per
year before the change), that is, an increased milk yield per cow of
21.5%. Herd J had a change of cowman after 4 years and this change
was associated with a massive reduction in yield of 29.2% (from an
average of 3100 to 2196 litres per cow). Herd L experienced 2 changes
of cowman after 2 and 4 years. The first change was associated with a

18

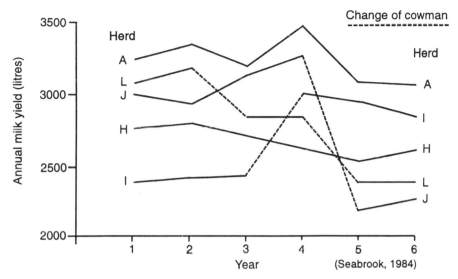

Figure 3.1   Milk production in single-man dairy herds

reduction in milk yield per cow from 3128 to 2835 (9.4%) while the second change saw a further reduction in yield over the next 2 years to 2368 litres per cow, that is, a 16.5% decrease.

Seabrook attempted to pinpoint the characteristics of high achievement cowmen in his study (see Table 3.2).

The indications from Table 3.2 were that cowmen who achieved the best results in terms of milk yield from their herds tended to be somewhat introverted, with a high degree of confidence in their own ability.

It is highly likely, of course, that desirable characteristics in a stockman who is the sole employee on a one-man unit are different from those needed in a stockperson who is a member of a team of workers on a large farm.

Seabrook (1984) also examined the behaviour of the cowman towards the cows in the parlour in 24 similar single-man dairy herds where feed inputs per cow were almost identical. His results are summarised in Table 3.3.

In the higher yielding herds the stockperson patted and stroked and talked to the cows far more frequently than did those stockpersons tending the lower yielding herds. This behaviour appeared

19

Table 3.2 Average characteristics of high-achievement cowmen (x)

	1	2	3	4	
Not easy-going	x				Easy-going
Adaptable			x		Unadaptable
Inconsiderate				x	Considerate
Meek				x	Not meek
Patient	x				Impatient
Unsociable	x				Sociable
Not modest	x				Modest
Independent minded	x				Not independent minded
Persevering	x				Giving up easily
A worrier			x		Not a worrier
Cheerful			x		Grumpy
Talkative				x	Not talkative
One who speaks one's mind		x			One who keeps quiet
Difficult to get on with		x			Easy to get on with
Lacking confidence				x	Confident
Cooperative				x	Uncooperative
Liking change				x	Suspicious of change
Forceful		x			Giving in easily

Source: Seabrook (1984)

Table 3.3 Stockperson behaviour towards cows in relation to milk yields in 24 one-person dairy herds

	Stockperson behaviour in parlour (number of times per minute per cow)	
	Touching cows	Talking to cows
Higher yielding herds	2.1	4.1
Lower yielding herds	0.5	0.6

Source: Seabrook (1984)

to increase the confidence of the cow in her attendant by improving the bond between them and this was reflected in other aspects of cow behaviour (see Table 3.4). Thus cows in the higher yielding herds entered the parlour for milking more readily and they dunged less frequently while in the parlour. They were less wary of the human observer, moving away a shorter (flight) distance when approached and approaching (perhaps investigating) an observer more frequently in a given time period. Seabrook suggested that this greater confidence in the stockperson, or better bond between cow and stockperson, helped to lessen stress in the cow in the parlour, put

Table 3.4   Behaviour of dairy cows on one-person dairy herds

	Higher yielding herds	Lower yielding herds
Mean entry time to parlour (secs/cow)	9.9	16.1
Field flight distance (metres)	0.5	2.5
Approaches to observer		
(number per min.)	10.2	3.0
Dunging in the parlour (number per hour)	3.0	18.2

Source: Seabrook (1984)

her more at ease and resulted in less inhibition to milk 'let-down' than in the lower yielding herds. The obvious implication was that differences in the efficiency of milk let-down in the parlour as influenced by the cow-stockperson relationship might well have been the major reason for the differences in performance between the lower and higher milk yielding herds.

2.   EVIDENCE FROM PIG PRODUCTION

Only a limited amount of well controlled scientific work has been carried out on aspects of stockmanship in pig production. The most comprehensive of such studies have been carried out by Dr Paul Hemsworth who is now based at the Werribee Research Institute in Australia. In earlier work in the Netherlands, Dr Hemsworth became involved in a situation with pigs similar to that of Martin Seabrook with the 12 single-man dairy herds under almost identical management in Britain. He had the opportunity to study 12 one-man-operated commercial sow herds under the one ownership and which were very similar in terms of location, buildings, genotypes, herd size, feeding and management advice. He found that, despite such similarities, reproductive performance varied very widely (see Table 3.5).

Table 3.5   Reproductive performance of 12 similar single-man sow herds in Holland

	Mean	Range	Advantage of best relative to worst
Total births per litter	10.4	9.4–11.3	18.9%
Farrowing rate (%)	86.4	79.5–94.4	18.7%
Total births per sow per year	20.6	17.9–22.5	25.7%

Source: Hemsworth *et al.* (1981)

It can be seen that the differences between the best and worst farms were very marked. In an attempt to explain these large between-farm differences, Hemsworth and his colleagues studied aspects of the behaviour of sows on these farms. Sows, when feeding, were tested for their withdrawal response to the hand of an experimenter lowered gradually to touch the side of the snout. Sows were also tested individually in a standard pen, in terms of the time taken to approach, and their interactions with, an experimenter. Large differences were found in behavioural responses between farms. Gilts and first litter sows appeared to be most wary of the experimenter in the 2 tests. However, when the results were corrected for parity differences, the farms on which sows appeared to have the best relationship with humans as measured by their behavioural responses were found to be the most productive in terms of piglets born per sow per year (as influenced mainly by farrowing rate). The behavioural responses towards the experimenter were found to be similar to those towards the stockman. It was suggested that on those farms where the relationship between the stockman and his pigs is poor, sows may show a stress response in the presence of the stockman, to the extent that reproductive failure may occur.

Thus Hemsworth and his colleagues found that in the best performing herds the sows were very much at ease with human beings, whereas sows on farms with poor performance were wary of humans. The implications were that on the high producing farms sows were handled a great deal by the stockman in a sensitive and friendly way, while on the farms where sows were much more wary and nervous of humans, the amount of handling that the sows received from the stockman was considerably less and/or the sensitivity and care of such handling was of dubious quality.

Hemsworth and his colleagues then proceeded to try to simulate caring and uncaring stockmanship in more controlled investigations. In one study they subjected 11 to 22 week old pigs penned in groups of eight to 'pleasant' and 'unpleasant' treatments three times per week over a period of two months. The 'pleasant' treatment consisted of entering the pen and stroking a pig when it approached while the unpleasant treatment consisted of entering the pen and slapping or giving a shock to the pig as it approached. At 25 weeks of age, the pigs in the unpleasant handling treatment were less willing to approach the experimenter when he entered the pen for 3 minute test periods. In addition, the pigs subjected to the 'unpleasant' treatment had a lower growth rate and had higher corticosteroid

Table 3.6   Effects of handling treatments on the level of fear of humans and performance of pigs in four experiments

Experiment	Handling treatment		
	Pleasant	Minimal*	Unpleasant
1. Hemsworth et al. (1981)			
Time to interact with experimenter (sec)†	119	–	157
Growth rate from 11–22 weeks (g/day)	709	–	669
Free corticosteroid concentrations (ng/ml)‡	2.1	–	3.1
2. Gonyou et al. (1986)			
Time to interact with experimenter (sec)†	73	81	147
Growth rate from 8–18 weeks (g/day)	897	888	837
3. Hemsworth et al. (1987)			
Time to interact with experimenter (sec)†	10	92	147
Growth rate from 7–13 weeks (g/day)	455	458	404
Free corticosteroid concentrations (ng/ml)‡	1.6	1.7	2.5
4. Hemsworth et al. (1986)			
Time to interact with experimenter (sec)†	48	96	120
Pregnancy rate of gilts (%)	88	57	33
Age of a fully co-ordinated mating response by boars (days)	161	176	193
Free corticosteroid concentrations (ng/ml)‡	1.7	1.8	2.4

* A treatment involving minimal human contact
† Standard test to assess level of fear of humans by pigs
‡ Blood samples remotely collected at hourly intervals from 0800 to 1700 h

Source: Hemsworth (1988)

concentration in the blood* both at rest and in response to the presence of the experimenter (see Table 3.6). It was concluded that the unpleasant handling by humans resulted in both chronic and acute stress responses. Other experiments with young pigs (Table 3.6) have since confirmed the earlier findings of Hemsworth and his colleagues.

In another experiment, similar 'pleasant' and 'unpleasant' treatments were imposed at a similar age by Hemsworth and his

* Selye (1976) postulated that exposure of animals, including man, to fear-provoking stimuli results in a range of physiological responses, one of the most consistent being elevated corticosteroid levels. Hemsworth and co-workers considered data on the corticosteroid response of pigs to humans to be useful in validating the use of behavioural data to assess the level of fear of humans by pigs.

associates on potential breeding stock, and a further 'control' treatment was examined in which stock had little contact with humans apart from that associated with routine husbandry practices. Relative to the 'pleasant' handling treatment, the control and 'unpleasant' treatments were associated with greater reluctance to approach humans, higher blood corticosteroid level following contact with humans, later sexual development and poorer conception rate to service at second oestrus (Table 3.6).

The results of the last experiment cited in Table 3.6 are particularly worthy of note. The 'pleasant' treatment was associated with a pregnancy rate of 88% whereas an extremely low pregnancy rate of 33% was associated with the 'unpleasant' treatment. The results of these controlled trials help to confirm the suspected reasons for the large between-herd differences in sow breeding performance found in Hemsworth's earlier studies in Holland. This indicates the great sensitivity and care which must be accorded to breeding stock in order to attain high levels of reproductive performance. The results of Hemsworth and his colleagues therefore provide scientific evidence that the behaviour of the pigman, the degree and nature of contact he has with his animals and the relationship he establishes with them, have an important influence on the welfare and productivity of the animals in his care.

Further relevant research work on sows and piglets has been carried out by Dryden and Seabrook (1986) in England. Their 'pleasant' handling treatments involved patting and touching sows and piglets regularly while their 'aversive' handling involved gentle hitting and slapping but not overt aggression. They found that their 'pleasant' handling treatment was associated with 7% higher daily liveweight gain in piglets between birth and weaning at 4 weeks of age (205 versus 192 grams per day). Seabrook and Darroch (1990), working in large commercial units under the one ownership in England, characterised stockpersons according to their personality (confidence and extroversion levels), aggression levels, the way they handled their pigs, their responses towards awkward animals, the way they drove their tractors and their interests outwith their work. They found some evidence that the stockpeople who applied more pleasant influences on sows and their litters achieved higher levels of survival in their piglets.

These researchers found that stockpersons having a low confidence score tended to react towards their animals in a more inconsistent and a more aggressive way e.g. using harsher methods when moving their animals. The aggressiveness of stockpersons or

Caring and sensitive stockmanship, in achieving better man–animal relationships, has a major influence on the milk yield in dairy herds and the reproductive performance of pigs.

the 'implied non-aggression level' of stockpersons was measured by an objective personality interview. Those stockpersons with higher non-aggression levels handled their pigs more gently.

Other important differences between pigs looked after by stockpersons with different personalities and non-aggression scores included the approach behaviour of the pigs to the stockperson, the incidence of stereotypic behaviour of sows and the relative excitability of sows when a person entered the house. Sows subjected to more touching and stroking behaviour by the stockperson were less excitable.

It is useful to have available at least some scientifically produced evidence to illustrate the important influence the stockperson can have, for better or worse, on the wellbeing and performance of our farm animals. Opinions on the importance of stockmanship in animal production are useful but scientific evidence of this impact is important in order to quantify the effect on the animal of each of the components of the influence of the animal attendant as well as the effect of the overall influence.

Assessing the economic impact of good, average or poor stockmanship on an animal production enterprise can be extremely difficult. The data presented earlier in this chapter showed the impact poor cowmen can have on milk output in dairy herds and the deleterious effects of aversive treatment by man on growth and reproductive performance in pigs. However, other practical effects of poor stockmanship have not been quantified. For example, what is the economic value of a timely diagnosis of a 'deviation from a healthy condition' in pig production followed by appropriate prompt treatment, relative to making such a diagnosis and administering treatment 3 or 4 days late? What is the economic value of an extra degree of interest, patience, commitment and dedication shown by a stockperson in charge of service or farrowing management in pig production? How much can a careless and uncommitted attitude of a stockperson cost when attending to feed mixing, administering feed to pigs, checking and amending the ventilation system, operating machinery on the farm or in administering drugs to treat the stock?

Perhaps it is less important to obtain an exact economic figure on the value of stockmanship than to be well aware of the potential importance and impact of stockmanship. Thus, while it is difficult to quantify the total economic effect of skilled stockmanship in pig production, an important objective at this stage is that there is full recognition that this factor has a major influence on technical and economic efficiency in animal production enterprises.

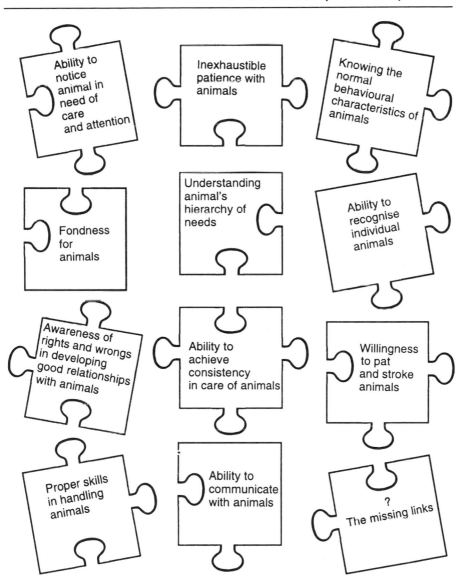

Figure 3.2 *The stockmanship puzzle! What are the desirable characteristics and how best do you develop them and fit them together within one individual?*

A fuller appreciation of the influences of the stockperson should encourage all those involved in managing animal production enterprises to pay much more attention than hitherto to selection of better stockpeople in the first place and thereafter to attend to relevant training and motivational procedures so as to progressively up-grade their influence for good on the animals in their care.

Summary

The characteristics of good stockmanship are summarised (as far as they can be at this stage of the book) in Figure 3.2. A more refined and complete definition will be synthesised as all relevant con-siderations pertaining to stockmanship are discussed in subsequent chapters.

As summarised in Figure 3.3, good stockmanship is important not only to the animal but to the stockman himself, to the executive of the business and to the community at large.

Thus, the animal and the community at large stand to gain from improving this most vital component of animal production systems.

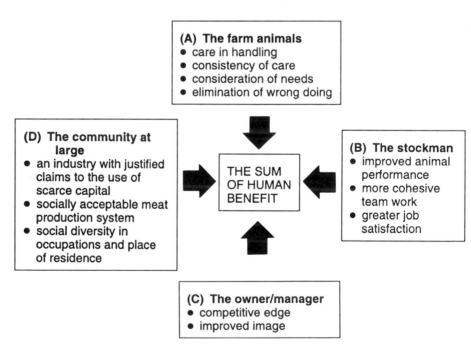

Figure 3.3   The importance of good stockmanship to the animal and to society

4

The role of empathy in stockmanship

Empathy is a key element of good stockmanship. Careful and gentle handling, together with effective communication with the animals, induces responses which have important influences on the health, welfare and productivity of the stock.

What is man–animal empathy?

Empathy describes the reciprocal relationship between man and the animals with which he associates, either by having them as pets or for the purposes of business, as in the case of farm animals. Empathy involves emotional attachment of man and animal and perhaps the best appreciation of most of what is involved in this mutual relationship is by owners of pets such as dogs and cats. Well cared for pets are attached to their minders not just because of 'cupboard love' but because they benefit emotionally from the relationship. In turn, communicating with and caring for pets has been demonstrated to fulfil needs in the human minders and be beneficial to them both emotionally and physiologically. 'Pet facilitated psychotherapy' has been used successfully to treat psychiatric problems and depression in children and adults by Mugford and McComisky (1974), Levinson (1978) and Corson and Corson (1980). In the latter study, uncommunicative patients became more attached to dogs as pets, the dogs appearing to act as social catalysts encouraging the forging of other relationships with fellow patients and staff. It was suggested that the humans became attached to the dogs because the dogs offered affection and reassurance without criticism. Thus, these and other studies and experiences show that the relationship between pets and their caring minders is a symbiotic one.

Homage to Rhuadh from Susie, Katie, Shona and Ewan. Well cared for pets are attached to their minders not just because of 'cupboard love' but because they benefit emotionally from the experience. In turn, communicating with and caring for pets has been demonstrated to fulfil needs in the human minders and be beneficial to them both emotionally and physiologically. Thus, the relationship between pets and their caring minders is a symbiotic one. (Photograph by Donald M. Mackay)

Empathy between man and animals kept for commercial purposes is more complicated due to:

- the larger number of animals in the relationship
- confinement of the animals
- economic pressures to be financially viable and profitable
- social pressures from individuals and groups critical of certain husbandry practices or opposed to any kind of husbandry of animals
- statutory obligations arising from legislation enacted as a result of social pressure

Man–animal empathy in general

Agriculture is not the only activity with animals kept for business purposes. Another economic activity which depends on animals is the circus. The situation of wild animals in circuses has been severely criticised from many quarters on the grounds that they are kept in alien, confined conditions and are trained to behave unnaturally to provide entertainment for circus goers. However, Dr Kiley-Worthington (1990), one of the world's foremost animal behaviour experts, carried out an independent scientific study of circus animals

An expert stockman has a wonderful ability to recognise the individual members of a large herd or flock and remember their particular eccentricities. The picture shows Peter Menzies (right) on his hill farm on the West Coast of Scotland. Peter runs 600 Blackface breeding ewes on his 1000 ha hill and breeds his own replacements. When he gathers the sheep and takes them to the sheep fold he can recognise each individual, he knows the heft (territory) on the hill to which they belong and he has the information on the mother and grandmother of each in the personal computer stored between his ears. In the confusion following gathering and penning he can also pair up each ewe and her lamb(s).

and came to the conclusion that, while there was room for improvement in terms of animal handling and care, circuses did not by their nature cause suffering and distress in animals.

The report of Kiley-Worthington emphasised the importance of good handling of the circus animals by their carers and trainers. She defined 'handling' as 'humans interacting with animals by touching, talking to, and being close to them over a period of time. The object of handling is fundamentally to reduce the fear that the animals may have of such close contact with humans and subsequently other unfamiliar objects. Once the animal can be handled easily and shows no sign of fear, as a rule both the animal and the human handler have begun to establish an emotional relationship which, if it progresses appropriately, can develop into affection and pleasure in each other's company. In addition, mutual levels of confidence with each other can build up to a point where the animal has confidence in its handler and will therefore go places and do things with him that it would otherwise be frightened to do.'

Many good stockpeople with experience of handling such farm livestock as horses, bulls, dairy cows, suckler cows, calves, beef cattle, sheep, goats, pigs and poultry will be able to identify with the above description of a strong relationship developing between man and the animals in his care and the attainment of mutual levels of confidence in each other.

Kiley-Worthington outlines some of the characteristics of good handlers of animals: 'Handling involves understanding the animal's body language but also being able to control one's own so that it does not portray certain emotions that one might be feeling, for example fear. Good handling must leave the animal with a pleasurable experience, otherwise the animal will quickly learn not to be handled. It is not difficult to teach most interested humans the first steps in good handling of animals but how this develops with experience in those humans depends on their interest, sensitivity and skills.

A key feature of good handling is that the animals and humans must have mutual respect. Insufficient respect by the human may result in the animal becoming more frightened of humans and sometimes attacking in defensive threat. Insufficient respect of the human by the animal will result in the human being completely manipulated by the animal and, as a result, the animal may become aggressive when it does not get its own way.'

Many stockpeople with experience of handling farm animals, and bulls and boars in particular, will appreciate the emphasis on the need for mutual respect between man and animal and also the need for the stockperson to conceal any fear through firm but sensitive handling.

In the proper handling and training of circus animals, Kiley-Worthington (1990) refers to the need to avoid negative reinforcement at all costs. Negative reinforcement refers to the animal being treated in some unpleasant way if it fails to do what the handler bids it. Such negative treatment is likely to lead to fear and/or aggression on the part of the animal and could damage relationships between trainer and animal irreparably. Positive reinforcement, that is, providing a reward when the animal conforms to the bidding of the handler, is effective in animal handling and training since this approach helps reinforce the developing good relationship between the handler and the animal.

The good stockperson on the farm has always been aware that, when encouraging a stubborn or nervous animal to move from A to B, it is easier to lead the animal in the desired direction by deploying the man-animal emotional bond along with an attractant such as feed and to duly reward it when it has completed the move rather than to cajole it or drive it from behind. The subtle rather than the aggressive approach always pays in such situations and helps to further cement the confidence of the animal in its attendant.

As well as the use of positive reinforcement in the form of food rewards in the training and handling of animals, Kiley-Worthington (1990) stresses the importance of 'praise with the voice.'

Stockpeople with experience of work with animals on small family farms will be able to relate to this emphasis on the importance of 'praise and the voice'. It used to be the norm to give each cow a name. This undoubtedly helped to overcome any self-conscious obstacle in talking to a cow! After all, the dog was called Ben and one talked to Ben. So it was natural enough to talk to a cow named Daisy. The naming of the animal when young—choosing a name for the newborn calf—gave a long term perspective to the expected contribution of the young animal to the business and the wellbeing of the family. It was worthwhile getting to know the animal and its behavioural characteristics and giving her an unmistakable identity—a name. This also contributed to error-free communication of information within the family: 'Keep an eye on Daisy—her milk was down yesterday' and 'Give Gertrude her feed while milking her, otherwise she will kick.'

It is possible that as the size of herds and flocks under the one stockperson has increased over recent times resulting in the blurring of identity of individual animals, and increased pressure on the attendant, he may have less time and inclination to talk to his animals and give them 'praise with the voice' as appropriate. Such conversing of the stockperson with his animals, for example when milking cows, feeding all types of livestock, testing for the presence of prefarrowing milk in sows by udder massage, herding stock and when harnessing, grooming and working horses, undoubtedly plays an important part in developing mutual respect and empathy and in increasing ease of handling.

Thus the 'talking to stock' and 'praise with the voice' components of the so called 'art' of stockmanship may well have suffered in the trend towards expansion of animal enterprises and the greater pressurisation of labour in these businesses.

Kiley-Worthington (1990) found that there are bad circus trainers around and advocated the establishment of a training school for all circus animal trainers. In the animal production industries there is a small minority of bad stockpeople and there is certainly no less need for training of personnel to staff our farm animal enterprises and to accord to the livestock therein the care and attention which is their

The photograph reflects not only the synchrony of the two Highland ponies, Dick and Joseph, but the communication skills and empathy of the ploughman, Roddy MacLeod, with his partners in the ploughing of seaside croft lands in Arnisdale in the West Highlands of Scotland.

Empathy between horse and rider was absolutely crucial when these units were deployed in a military context.

due. Without this training in the animal production industries there will continue to be an element of ill caring of animals while stockmanship in general will not attain the level of excellence which is absolutely essential for high standards of animal health, welfare and productivity.

Man–animal empathy in agriculture

The existence and influence of empathy between stockperson and farm animal is generally, although not universally, appreciated in agriculture. Some authorities in trying to define 'good stockmanship' consider that it consists solely of using technically correct methods in handling and managing farm livestock. The impact of the emotional relationship between stockperson and animal is therefore largely discounted in such definitions.

However, other students of stockmanship show a much fuller appreciation of the important role of empathy between stockperson and animal. Seabrook (1982) stated that the stockperson in tending the farm animals in his care may assume one or other of the following roles: boss animal, mother substitute, leader or friend. Provided that the term 'boss animal' implies consistently firm but sensitive control of his charges, all these terms imply mutual emotional attachment.

The ability of different stockpeople to achieve widely different performance from similar resources in terms of livestock, feed, housing and all other requirements has been highlighted by Lloyd (1975): 'From experience it is appreciated by poultry managements that a change of poultryman during a production programme can produce dramatic and permanent changes for better or worse.' However, probably alluding to that influential but unquantifiable component of the rapport or relationship between the stockman and his charges, he proceeds: 'but there is little information available which may help to predict or explain the extent of such changes.'

Those changes made in farm businesses which have fairly predictable results are mentioned by Lloyd (1975): 'Economists have long established the difference which management skills can make to business results on farms of similar type. The effects of changes in management policy and husbandry practices (e.g. feeding levels, production timing, temperature and ventilation) are all easily demonstrated and repeatable under experimental conditions. Thus the bearing which proper accomplishment of specific tasks has on results, is generally well known and frequently quantified.'

Proceeding from the influences of the stockman which are quantifiable, Lloyd (1975) then addresses the 'additional variable human factor which has a marked effect on results achieved from animals under human care. This variable factor is referred to by some practitioners as the 'art of stockmanship'. It enables some individuals to achieve a response from animals which cannot be achieved by others using the same resources and techniques. It is largely imbued in a person through inheritance and experience although the indications are that the characteristics can be learned to some extent by others.'

In attempting to explain why stockmanship does 'not consist solely of using technically correct methods' because 'it has proved impossible to identify the total detail of technically perfect methods so that they are repeatably reproducible by all staff', Lloyd cites many findings from non-farm animal species to help explain why 'the stockman conducting the task usually adds some special and individual characteristic to the work method.'

The general agreement 'in medical circles that human babies subject to mothering and cuddling become better orientated psychologically and may also develop more satisfactorily physiologically and enjoy better health than those reared in a technically correct environment lacking contact comfort' is acknowledged. Lloyd also cites evidence from avian and mammalian studies of the importance of a mother-figure with young animals from early life on their subsequent behaviour and emotional stability in group situations in later life. This early bonding between the newborn and the mother is transferable to a mother substitute such as a stockperson with sympathetic and caring qualities.

Extrapolating from such evidence with other species, Lloyd states 'it is a reasonable assumption that the closeness and form of the relationship between an agricultural stockman and the animals in his charge will affect their state of emotional security and behaviour. For those species where emotional security is linked with agricultural productivity, the relationship between the stockman and the animals in his charge is important.'

The example used by Lloyd to illustrate such an influence of the stockperson on the animals in his care was that of the cowmen in single-man dairy herds reported by Seabrook (1974) and cited earlier, in Chapter 3. The influence of the individual cowmen in obtaining widely divergent milk yields from their virtually identical herds (see Tables 3.1 and 3.2) was claimed mainly to derive 'from the personality type of the stockman, for this leads to predictable forms of behaviour in the association between the man and his cows—degrees of firmness, kindness and sympathy, "fellowship" and constancy of behaviour. Cows respond through placidity and confidence by the full release of milk and possibly an improvement in physiological and metabolic processes.'

Although Lloyd is careful to point out that the possible effect of the stockman on 'physiological and metabolic processes in the cow' could not be supported by available scientific evidence at that time, later research and particularly that cited in Chapter 3 by Paul Hemsworth and his colleagues working with pigs would fully vindicate these hypotheses regarding the full impact of the stockman's influence as suggested by Lloyd.

The 'aversive' treatments used by Hemsworth and his colleagues involved either slapping or briefly shocking the pig with a battery-operated prodder whenever the pig approached the attendant. 'Pleasant' treatments involved either patting or stroking the pig whenever it approached the attendant. These handling treatments therefore varied from treatments which discouraged approach to humans, by using aversive (negative nature) handling whenever the pig approached, to treatments which encouraged approach to humans, by using apparently pleasant handling whenever the pig approached the experimenter.

The results indicated that the aversive handling treatments increased

The early bonding between the newborn and the mother is transferable to a mother substitute such as a stockperson with sympathetic and caring qualities if the need arises, e.g. if the ewe's milk production fails.

the level of fear of humans, increased corticosteroid levels and depressed growth rate in young pigs and reproductive performance in both gilts and young boars.

Thus, simulated poor stockmanship had an adverse effect on the stockman-animal relationship and resulted in poorer pig welfare and performance. Hemsworth and his fellow researchers attempted to isolate some of the behavioural and non-behavioural characteristics of humans which are conducive to the development and maintenance of good human-animal relationships, in which the animals are not fearful of humans. They found that 8 to 12 week old pigs developed a close relationship with a 'stockman' more quickly when he did not approach the pig, when he squatted in the pen, had bare hands and did not initiate interactions with the pig, than when he did approach the pig, stood erect in the pen, had gloved hands, or when he initiated the interactions with the pig.

In particular, this study gives some indication of the human signals which the pig perceives as threatening. However, much more research is required to determine the nature and type of signals (visual, auditory, olfactory or tactile) which are most conducive to the rapid attainment and maintenance of a strong stockman-animal bond.

In view of the importance of avoiding actions or postures perceived as threatening by the stock if close man-animal bonds are to be developed, the oft quoted posture of the stockman of old, leaning on the fence or pen wall for some time to examine his stock to check wellbeing, may be relevant. Such a static posture does attract the stock perhaps through inquisitiveness but, if the stock approach the attendant voluntarily in this way, the man-animal bond is likely to be developed.

Neither is there any threat to the animal, but rather the reverse influence, in leading stock rather than driving them when moving between two locations. In many parts of the world, sheep and goats are still handled in this way as of old, the words of Psalm 23 describing the handling of the 'sheep' by the Good Shepherd:

'He lets me rest in fields of green grass
and leads me to quiet pools of fresh water.'

It is likely, on the basis of experience with children and pets, that the strength and value of the man-animal bond is enhanced by developing it as early in the life of the animal as possible.

Hemsworth and his colleagues examined this aspect by imposing treatments differing in the type and frequency of handling of young pigs in the first 8 weeks of life. They concluded that careful and frequent early handling of pigs can be beneficial in creating desirable

The dedicated stockperson who spends time leaning on the fence or pen wall examining his stock often gets accused by the boss of wasting precious time. However, such time can be very well spent if the early stages of a problem in one or more animals can be detected and treated there and then before it becomes more serious. In addition, such a static posture attracts the stock, perhaps through inquisitiveness, and in the voluntary approach to the attendant in this way, the man–animal bond is likely to be developed.

behavioural responses to humans later in life.

Thus this 'empathy' component of the 'art' of stockmanship must be recognised as being very influential. It appears to be influenced by the personality type of the stockman and, according to Lloyd, personality characteristics in turn 'are largely a product of inheritance, upbringing and social experience'. Thus part of this empathy is in-built at birth by virtue of inheritance. However, the oft quoted statement that 'good stockmen are born, not made' is likely to be only partially true for it appears that empathy between man and the animals in his care can be influenced by the upbringing of the man and by his experience. Since education and training contribute to upbringing and experience it is vitally important to consider ways

Production system	Size of unit/work organisation
The production system and/or building design and/or the provision of automation reduces the opportunity for close and regular man-animal contact and is thus not conducive to the development of effective stockman-animal empathy.	Scale of production so big and workforce organisation is such that individual stockmen do not have much opportunity to get to know and develop a relationship with each animal.

EMPATHY POORLY DEVELOPED

Management	Education and training
• Replacement stock obtained from source where stock are not handled a great deal and are sometimes subjected to insensitive and unpleasant treatment. • The farm is understaffed so that stockmen have insufficient time to devote to individual animals and to build up a good relationship with the animals.	• Relatively uneducated stockpeople recruited and not evaluated in terms of empathy with animals. • Inadequate training provision to educate stockpeople in the needs of the pig, pig behaviour and pig body language. • Inadequate training on pig handling skills conducive to development of good empathy with the animals due to overemphasis on the technical operation of the production system.

Figure 4.1   Management-related factors which can impair stockman–animal empathy

41

of measuring 'empathy' when recruiting stockpeople in the first place and then trying to develop this most valuable characteristic in stockpeople more strongly by appropriate training and by providing the most appropriate working environment which will be conducive to allowing the full expression of this quality in stockpersons towards the animals in their care.

Summary

We can now sum up the major influences on empathy in animal production. Figure 4.1 identifies the management related factors which, if allowed to develop, can impair stockman-animal empathy. Figure 4.2 summarises the factors that are likely to reinforce positively the degree of empathy which exists between stockman and animal.

It is clear that education and training as well as provision of a good working environment are influential in the development and maintenance of empathy in pig production systems. Appropriate training of stockpeople and provision of a stimulating working environment form the basis of subsequent chapters.

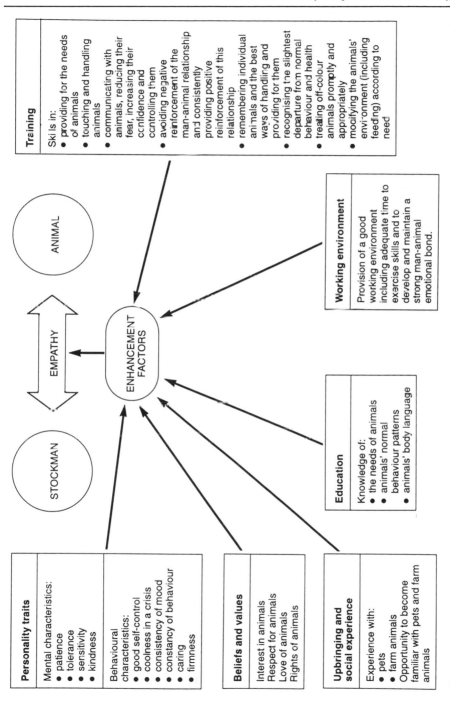

Training

Skills in:
- providing for the needs of animals
- touching and handling animals
- communicating with animals, reducing their fear, increasing their confidence and controlling them
- avoiding negative reinforcement of the man-animal relationship and consistently providing positive reinforcement of this relationship
- remembering individual animals and the best ways of handling and providing for them
- recognising the slightest departure from normal behaviour and health
- treating off-colour animals promptly and appropriately
- modifying the animals' environment (including feeding) according to need

ANIMAL

EMPATHY

ENHANCEMENT FACTORS

STOCKMAN

Working environment

Provision of a good working environment including adequate time to exercise skills and to develop and maintain a strong man-animal emotional bond.

Education

Knowledge of:
- the needs of animals
- animals' normal behaviour patterns
- animals' body language

Personality traits

Mental characteristics:
- patience
- tolerance
- sensitivity
- kindness

Behavioural characteristics:
- good self-control
- coolness in a crisis
- consistency of mood
- constancy of behaviour
- caring
- firmness

Beliefs and values

Interest in animals
Respect for animals
Love of animals
Rights of animals

Upbringing and social experience

Experience with:
- pets
- farm animals
Opportunity to become familiar with pets and farm animals

Figure 4.2   Factors conducive to the development of excellent empathy in animal systems

43

5

Factors influencing the quality of the stockperson

Good stockmanship is dependent on the personality traits and attitude of the individual, on affinity with animals, on appropriate education and training, on job satisfaction and motivation.

If those who appreciate the importance of 'stockmanship' were asked to list the most essential qualities of a good stockperson, the terms 'interest in and respect for animals', 'patience', 'tolerance', 'good observational or perceptual skills' and 'ability to pay attention to detail' would be cited frequently.

One could generalise by stating that good stockmanship is displayed by the person who has the appropriate behaviour towards the stock, combined with the motivation, skills and experience to deal, in the best possible way, with all aspects of the job of tending to the animals in his care.

Having the correct attitude and behaviour depends greatly on the basic nature and personality of the person but few authorities would deny that these characteristics are partly learned—either consciously or unconsciously. Such learning can come about through personal experience and/or positive training in understanding and skills.

Thus, attitudes towards stock in animal production systems can be influenced by colleagues, through work experience and by management policy through example, instruction and training. These influences can be very important in inducing or instilling the appropriate attitude of stockpeople to the animals in their care.

The basic attitudes of stockpeople towards their animals can also be influenced by negative factors which have an adverse effect on job satisfaction and by positive attitudes towards motivation of the workforce.

Thus the factors which influence the overall qualities of the stockperson can be summarised as follows:

- Basic interest in and respect for animals
- General attitude in tending and handling animals (care and diligence)
- Amount of experience in looking after stock
- Training in understanding the animal and its needs
- Training in animal caring skills
- Removing sources of dissatisfaction
- Motivating influences

Some of these influences are worthy of further comment at this stage.

Experience

Experience can be regarded as the product of time spent working in animal enterprises and the knowledge and understanding acquired. In pig farms which have a high turnover of staff, inexperienced new staff could be entering the system regularly to replace departing personnel. The consequent short-falls in experience among the staff are likely to be a severe disadvantage to animals and the enterprise in many respects. In such a situation, urgent steps must be taken to determine the basic reasons for high staff turnover. Once the causes of the problem are detected, the most appropriate corrective action must be taken.

Skills

Regarding the perceptual and technical skills required to be a good pig stockperson, the basic elements of most skills can be passed on through simple instruction and training sessions. Such procedures as stock handling and control, oestrus detection, management of natural services or of artificial insemination, monitoring farrowing, recording rectal temperature, providing strategic assistance to farrowing sows or newborn piglets, fostering, injecting, feeding and general stock observation are all relatively simple to learn provided the basic interest and attitude of the pupil are appropriate and the instruction is carried out by competent, well informed people with patience and clarity. Instruction at a higher level to increase understanding of the anatomy and physiological processes of the animal will be of great value because such increased awareness will make the stockperson more sensitive to the needs of each animal in his care and will help him to adapt his caring not only to the differing needs of individuals but to make appropriate adjustments in approach to deal with changes in environment or management system.

Sources of dissatisfaction or negative influences on the stockperson

Several authorities have drawn attention to factors associated with animal enterprises which have negative influences on the stockpeople employed and have adverse effects on their attitudes to work including the quality of care of their animals.

Seabrook (1982) suggested that employment in some intensive animal production systems was likely to modify the normal behaviour of stockpeople and that the degree to which each was adversely affected would be dependent on personality, attitude, experience and training.

The following are some of the ways in which the behaviour of stockpeople in their work can be adversely affected.

HABITUATION

In habituation a stockperson becomes accustomed to a system and accepts it for what it is. This may mean that he accepts mistreatment of the animals without challenge but it also may mean that he will resist any change.

PERCEPTION ADAPTATION

When signals (e.g. from animals in his care) arrive at the brain of the stockperson they are processed and interpreted. This interpretation (or perception) by the stockperson is influenced by many factors including past experience and the stockperson's own attitude and characteristics. In part, the stockperson may 'see what he wants to see' and not what is actually there; this may mean that aspects of the behaviour of the animal are interpreted relative to the stockperson's own perception rather than in a totally objective manner. Obviously, if the stockperson wrongly interprets the animal's signals through its behaviour he may fail to recognise a problem in the animal and will therefore not be responsive to its needs.

SELECTIVE ATTENTION

The stockperson receives a range of visual, noise and smell signals from the stock with which he works and from the general environment round about. However, he often receives more signals than he can respond to. He deals with overload of signals by selecting out only those signals which he feels are important. For example, the skilful stockperson will distinguish (and respond to) the squeal of a piglet that is being overlain by the sow from the squeals of litter-mates as they fight competing with each other for teats at milk 'let-down' periods. However, the stockman is also able to modify this selection process of the signals to which he will respond by raising his threshold level for responding to signals, which results in him paying no attention whatsoever to certain noises. When he becomes less responsive to animal signals in this way he may fail to detect critical problems among his stock. A response analogous to this which many pig stockpeople recognise is that applied when many sows in a building are being hand fed. There is a great collective high pitched screaming noise from the sows in the excitement at their anticipation of being fed and the stockperson appears to have an ability to dull his sense of hearing as a defence mechanism until all sows have been fed and peace reigns once again.

LACK OF IDENTIFICATION

Many intensive systems are well managed and this helps the stock-people to identify with them. However, some stockpeople, particularly on less well managed enterprises, become alienated from the system and fail to identify with the tasks, the animals and the objectives of management. Lloyd (1975) has referred to this problem in intensive poultry production and the tendency in that industry towards 'organisational rather than biopsychological policies in the quest to improve labour efficiency. Thus, today, labour economy is most usually approached by minimising the use of the costly labour resource through mechanisation and automation or through the economics of large scale operation. Some of the resulting working conditions tend to discourage the man with a "feel" for poultry. He may have to endure dusty and dirty working conditions in buildings devoid of natural light. He is overwhelmed by sheer numbers of birds and many of the jobs which involve his skill and provide satisfaction and fulfilment are excluded from his work by automation and mechanisation. Too often he is an automaton.'

The above and other influences can induce negative attitudes to work like boredom, frustration and alienation, and Bennett (1989) writes about these as follows:

BOREDOM

Boredom may result from continuous repetition of a simple task, or from the social environment in which tasks are undertaken. A task might be interesting, but the worker still feels bored if he or she must complete it in isolation. Equally, jobs can be trivial and repetitious, yet not create boredom because workers are able to communicate pleasurably with others. Workers who perform complex and challenging tasks typically become absorbed by them and are not bored. A good example of such tasks is the supervision of and provision of intensive care at lambing or farrowing. It takes a great deal of experience and behavioural monitoring to know when the farrowing sow needs assistance. In addition, when several sows are farrowing simultaneously and large litters are being born a great deal of careful thought and skilled actions are necessary in arranging for smaller

Sheep shearing, wool rolling and wool packing in operation on a hill sheep farm in the Highlands of Scotland. It is most important to avoid the negative influence of boredom in the work organisation of stockpeople. A task like sheep shearing might be interesting but the worker can still feel bored if he or she must complete it in isolation. However, when working with a group of like-minded people, there is lively communication, relating to stories from the past, guidance (training) of the young and a great deal of light-hearted humour. This changes the whole atmosphere of being lonely and boring to a lively, pleasurable, and stimulating experience. (Reproduced by kind permission of National Museums of Scotland)

Jobs can be trivial and repetitious, yet not create boredom because workers are able to communicate pleasurably with each other. The only connection of the turnip singling activity depicted in this photograph with animal production is that sheep, cattle and horses are the recipients of the mature crop. The boredom of working on one's own in a large field bears no resemblance whatsoever to the fun the job can be in a good squad where the stories, leg pulling and the rapport in general can provide a high level of entertainment. (Reproduced by kind permission of National Museums of Scotland)

piglets to obtain their due share of colostrum and then to arrange fostering and crossfostering in the best interests of all the piglets and the small ones in particular. To the dedicated stockperson this is certainly a very absorbing activity and there is no risk whatsoever of boredom setting in.

FRUSTRATION

Workers experience frustration when they are prevented from exercising control over their work and are not able to achieve their (self-defined) objectives. Frustration can be caused by the lack of control over working methods or the speed of production, by having to do work perceived as meaningless, through not being involved in decision making or through feeling that individual grievances have

49

not been properly heard. A worker may react to frustration positively, by attempting to overcome the problem that caused the obstruction, or negatively. Examples of negative reactions are aggression (quarrels with colleagues, hostility towards management), apathy (lateness in arriving at work, absenteeism), unwillingness to assume responsibility, poor quality work, high propensities to have accidents and high rates of labour turnover.

ALIENATION

This is a feeling that work is not a relevant or important part of one's life; that one does not really belong to the work community. It is associated with feelings of discontent, isolation and futility. Alienated workers perceive themselves as powerless and dominated. Work becomes simply a means to achieve material ends. Great unhappiness can result from alienation; indeed, the mental or physical health of the employee can suffer. Alienation may result from lack of contact with other workers and/or with management, from authoritarian or paternalistic management styles or simply through the boredom of routine work. Its consequences are numerous: poor quality output, absenteeism, resistance to change, industrial disputes and deteriorating interpersonal relationships.

People usually work better when they feel they have a personal stake in the success of their activities—success not necessarily being measured in financial terms.

Summary

The major factors influencing the quality of the stockperson are presented in Figure 5.1.

DEALING WITH SOURCES OF DISSATISFACTION

Because negative influences on the behaviour and attitudes of the stockman have such far-reaching effects on his abilities to tend the needs of the animals in his care and on his job satisfaction, management must be fully aware of such negative influences and do everything possible to eliminate them. Approaches towards achieving this objective form the basis of Chapter 9.

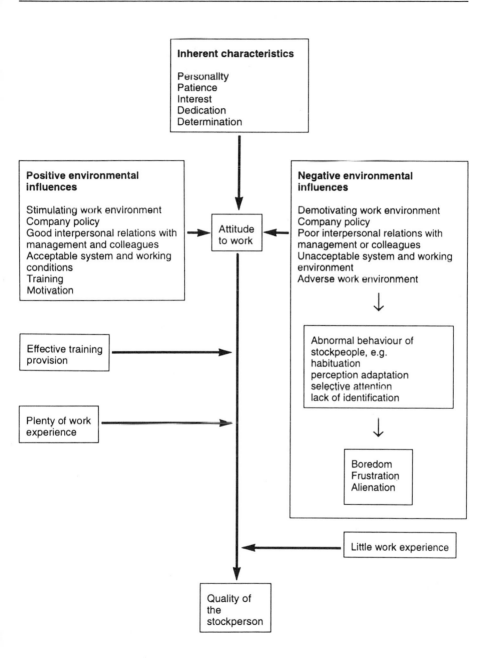

Figure 5.1   *A simplified model of the factors influencing the quality of the stockperson*

MOTIVATING STOCKPEOPLE

Removing negative influences on stockpeople does not constitute positive motivation. However, once such negative influences are eliminated, management is then in a position to make the most of motivational influences to get the best out of potentially good stockpeople in terms of effective care of their stock and the improved performance in physical and financial terms of the livestock enterprise. Approaches towards providing adequate motivation for stockpeople are dealt with in Chapter 10.

6

Attracting applicants and selecting the best

When you select the person you select a potential contribution to profit. (Freemantle 1985)

Attracting all those who want to work with farm animals to a vacant stockperson post must be given a high priority by management in the pig industry. The careful and objective selection of the best candidate from the full list of applicants will go a long way towards ensuring the commercial viability of the production unit and the competitiveness of the pig industry. These are the two assertions which form the foundations of this chapter.

The chapter contents are slanted towards the manager with responsibility for recruiting and selecting a stockperson. The recruitment stage seeks to attract those who are keen to work with pigs and the means to this end include the job description and its promotion/advertising. The selection stage seeks to pick the best person for husbanding the pigs and the one who will also fit nicely into the existing team of stockpeople. The means to this end include the assessment of candidates and the choice of the candidate considered the most suitable for the job. These are crucial management actions and decisions in the operation of a successful pig enterprise.

However, the interests of the stockperson are not ignored in this chapter. For the stockperson, the chapter provides a useful insight to good practice in the recruitment and selection of a stockman. Without this awareness, the candidate for a stockperson post is likely to find the recruitment and selection procedure somewhat baffling and stressful.

Well-thought-out and carefully conducted procedures are in everybody's interest. The employer, the employee and the pigs themselves will all benefit from the appointment of the right person for the job.

The main reasons for recruiting new staff are:

- an increase in the work load of the unit; for example, due to the implementation of new welfare codes
- the expansion of the business
- the departure of a stockman due to resignation or retirement
- the dismissal of a stockman; for example, for maltreatment of the pigs

Recruitment needs should be anticipated as far as possible in advance in order to allow a proper selection scheme to be implemented.

The first consideration in the appointment process is whether there exist internal candidates for the vacant post. This aspect is especially relevant when the vacancy is for a senior stockperson. Bennett (1989) makes the important point that much time and expense can be saved through internal promotions or transfers. Moreover, a policy of internal promotions enhances the morale of currently employed staff. However, even if a senior post is filled by an internal promotion, a replacement will be required for the job done previously by the promoted member of staff.

Factors influencing the number of applicants

The number of potential candidates attracted to a job vacancy is dependent on the following factors:

- number of people seeking work in the local labour market
- degree of competition for workers in the local labour market
- the appeal of the job
- the image of the industry
- attraction of the farm
- extent of promotion and advertising of the post
- scope for job flexibility
- wages
- promotion prospects

These will now be discussed in turn.

NUMBER OF PEOPLE SEEKING WORK
IN THE LABOUR MARKET

When labour is in short supply the task of attracting a reasonable number of applicants for a job vacancy is much harder than when considerable unemployment creates a strong flow of new entrants into the labour market. At any one time, the potential stockperson must be attracted from a labour pool consisting of:

- those persons already in work but considering a change of occupation
  This is what economists refer to as 'frictional change in un-employment'. Also, in the case of stockmanship, there may be a change in lifestyle involved e.g. when someone presently living and working in an urban environment is being targeted.
- those persons presently in work who have already been told, or can anticipate, that redundancy is imminent
  This is what economists refer to as 'structural change in unemploy-ment'. The firm that employs them may have been losing its market, and practices such as shorter working days, shorter work-ing week and prolonged holiday period are signals of market forces which in most cases lead to eventual closure of the business.
- those persons already without work who are able and willing to seek work
- those persons re-entering the labour market
  This component of the pool will consist predominantly of women who have been out of the labour market while having and raising their children.
- those persons who are entering the labour market for the first time
  The number of this component of the available stock of labour depends on the past pattern of births. For example, demographic statistics indicate that in Britain the supply of young recruits for employment will contract during the 1990s due to a low birth rate in the 1970s (see Figures 6.1 and 6.2). This demographic 'time bomb', as several authorities term it, is summarised by Curnow (1989) in the following terms: 'the number of 16 to 24 year olds is to fall by 1.2 million within the next 5 years (1990–95) i.e. a decline of 20%, while the reduction in 16 to 19 year olds within the same period will be of the order of 23%.'

Projections by the Government Actuary's Department suggest that the population of working age will increase by only 550,000 during

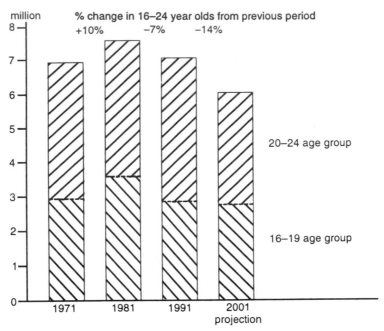

Figure 6.1 *Estimates and projection of the resident population in age groups 16 to 19 and 20 to 24 of Great Britain, 1971–2001*

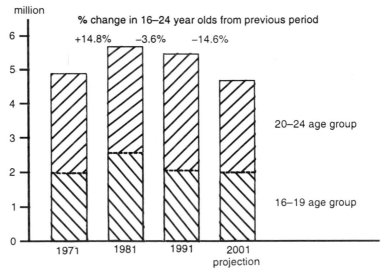

Figure 6.2 *Estimates and projection of the civilian labour force in age groups 16 to 19 and 20 to 24 of Great Britain, 1971–2001*

the 1990s compared with a rise of 1,750,000 in the 1980s. Almost all the projected increase in the civilian labour force during the 1990s is among women.

For the age group 16–24, the total in the civilian labour force will contract by an estimated 14.6% between 1991 and 2001 relative to the previous decade (see Figures 6.1, 6.2 and 6.3). The labour force in 2001 will be older, on average, than in 1991 due to a fall of around one million persons aged under 25 and a rise of 1.6 million people aged 25–54 in the labour force.

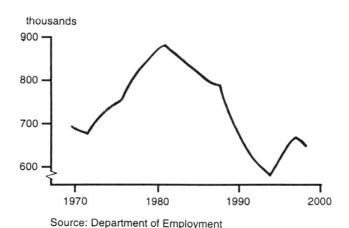

Source: Department of Employment

Figure 6.3   Persons at age 17 in Great Britain

DEGREE OF COMPETITION FOR WORKERS IN THE LOCAL LABOUR MARKET

The demand for labour is dependent on the buoyancy of industry in general and of those sectors with high labour needs in particular; this in turn is influenced by factors such as interest rates and the general level of wages.

THE APPEAL OF THE JOB AND THE IMAGE OF THE INDUSTRY SECTOR

The more attractive the image of an industry and the status of a job the higher the number of potential recruits that will be attracted to apply for posts.

In this context, the pig industry would appear to have a poor image. Goodman (1990) points out that only a very small minority of the general public in Britain have much understanding of what is involved in working on a pig farm. She points out that the type of media attention focused on the industry in the context of such issues as animal welfare, and also pollution associated with effluent disposal, portrays it in a bad light. She also reported the results of a small survey carried out by a pig breeding/processing company in England which had the objective of finding out the views of a class of forty-six 15 to 16 year old schoolchildren on the pig industry. They were asked if they would like to work on a pig farm and, if not, to state their reasons. As many as 34 out of the 46 (74%) responded 'no', only 10 would consider it while 2 were not sure. The reasons given from a prepared list of 11 possibilities for not wishing to work on a pig farm are summarised in Table 6.1.

Thus, the main reason for not wanting to work on a pig farm was dissatisfaction with (their awareness of) the conditions in which pigs are kept. Their regular use of terms such as 'cramped', 'cooped-up' and 'crammed together' suggested to Goodman a media influence on the opinions of the youngsters.

Table 6.1 Reasons given by 15 and 16 year old schoolchildren for not wanting to work on a pig farm

Reason	Percentage giving this reason
Conditions pigs are kept in	47
Pigs get killed	32
Do not like pigs/animals	29
Unpleasant environment	29
Concerns about the poor opinions of others about a job on a pig farm	20
Low pay	18
Other career in mind	18
Respondent too intelligent	15
Vegetarian	12
Do not know what is involved	9
Mundane, boring	3

Source: Goodman (1990)

ATTRACTION OF THE FARM

Perhaps somewhat surprisingly, potential employees are attracted more to large than to small businesses; in view of the fact that the average number of employees per pig farm tends to be small, this makes the industry much less attractive to many. In their studies in Ontario in 1990, Howard and his colleagues found that more applicants were attracted to job vacancies on large relative to small pig farms and, while the larger farms did not necessarily offer higher wages, they did have more benefits and potential for promotion. Goodman cites several possible attractions of larger farms:

- a well defined and impressive career structure
- comprehensive in-house training programmes
- better working conditions i.e. canteens and personal hygiene facilities
- a sense of belonging to a team
- they may employ full time personnel managers who can commit themselves to human resource issues

While some individuals may be more attracted to a job on a small pig farm because of an anticipated closer contact with the manager/farmer and greater involvement in the full range of activities including decision making, the priorities of a majority of potential recruits would appear to be different.

PROMOTION AND ADVERTISING

It is most important that the attractions of a career in the pig industry are publicised more effectively and that vacancies are also effectively advertised. Goodman (1990) has suggested that school children should be made more aware of opportunities in the pig industry before they begin to make their career choices. She cites the examples of other industries which make children more aware of career opportunities in their particular sector by producing, in a very professional manner, attractive project material which is included in the children's school curriculum. This allows children to learn about the particular industry and at the same time to develop understanding and skills in basic subjects which are a standard part of their education. For example, in studying project material relating to the pig industry, the children can develop understanding in arithmetic, mathematics, biology, physics, chemistry, economics and environmental sciences.

The amount of caring and nursing which has to be exercised in the successful management of a pig enterprise, particularly in the farrowing quarters, could be highlighted more in attempting to increase the appeal of the industry to prospective employees. The pressing need in the pig industry for personnel who have a great interest in and affection for animals, and who are able and willing to develop a mutually rewarding relationship or empathy with them, should also

The agricultural industry should make schoolchildren more aware of their sector and of relevant career opportunities by producing, in a very professional way, attractive project material for inclusion in the children's school curriculum. This allows the children to learn about agriculture and animal production and at the same time to develop understanding and skills in basic subjects which are a standard part of their education. For example, in studying project material relating to the pig, dairy or sheep industries, the children can develop understanding in arithmetic, mathematics, biology, physics, chemistry, economics and environmental sciences.

The amount of caring and nursing which has to be carried out in the successful management of a pig enterprise, particularly in the farrowing quarters, could be highlighted more in attempting to increase the appeal of the industry to prospective employees.

be emphasised. Such a portrayal could well attract pet lovers and owners to the industry and stimulate their interest, e.g. in becoming farrowing house attendants, even though they have no connection with agriculture. Having been recruited to the farrowing section and mastered sow and piglet management there, they then might not be unwilling to progress to work and gain experience in other sections of the pig unit.

Levinson (1978) claimed that experience of caring for a pet during childhood could make a person more sensitive to the feelings and attitudes of others and could inculcate tolerance, self acceptance and self control. It is possible that this experience and the qualities induced by it might well also make such a person more suited to a job as a stockperson in later life.

Regarding the possibility that pet lovers in general constitute a

Mutual trust in each other's company—a budding stockman in the making if ever there was one. (News Productions. CH-1446 Baulmes)

useful potential source of recruits to the pig industry, a recent study at the University of Aberdeen (English, Burgess, Bell and associates 1992) examined the possible association between experience of and attachment to pets and the abilities and attitudes of 25 pig stockpeople (21 male, 4 female) on 4 commercial pig farms. 92% of the stockpeople had experience of pets as children and all enjoyed that experience, while 72% had at least one family pet at present. Of 10 reasons listed for taking a job with pigs, 'liking pigs' or 'liking animals' appeared among the 3 most influential factors in 92 per cent of the stockpeople. All 25 stockpeople were in the habit of 'talking to' their pigs while 10 'stroked or rubbed' their pigs in friendly ways regularly in the course of their duties. The managers on the farms were asked to score each stockperson on a 1 to 10 scale (10 = perfection) for (A) overall competence as a stockperson, (B) 17 individual stockmanship characteristics and (C) 7 empathy related characteristics. Mean score for criteria A, B and C was 6.6, 6.8 and 6.8 respectively. Those 7 stockpeople for whom 'liking pigs' or 'liking animals' was their main reason for taking a job with pigs had higher overall (A) and 'empathy' (C) scores. The 4 females obtained higher scores in all 3 sets of criteria than the 21 male stockpeople. Those 11 stockpeople who believed that their experience with pets helped them in looking after their pigs also obtained higher mean scores in criteria A, B and C than the other 14. One of the many interesting aspects emerging from this study was the very genuine affection the stockpeople had for pets and for their pigs. This is a most valuable quality which must be deployed more effectively through education and training and ensuring good job satisfaction and motivation.

The pig industry in Britain has recently become much more aware of the need to sell itself to prospective employees. Displays at agricultural shows and career conventions, invitations to parties of school children to visit well operated pig farms and production of attractive publicity material in the form of glossy literature and videos all help to portray the pig industry in a more attractive and realistic light than that conveyed by much of the main media sources.

THE JOB DESCRIPTION

An essential prelude to advertising a stockperson job vacancy is to draw up a job specification. Drawing up a job description is a very useful exercise that will not only help create the most appropriate job vacancy advertisement but will also increase the understanding of the manager regarding the type of person he desires for the job. The text must be clear and simple and must never promise conditions or promotions that are unlikely to materialise, since such false promises will quickly lead to frustration on the part of the new appointee. Cleary (1990) outlines the following purposes of the job specification:

- it helps define the capabilities necessary in the successful applicant to fulfil the job functions
- it can be used as a selection aid during the job interview
- it prevents the disappointments which can occur when the employer claims that the employee is not adequately fulfilling an expected function while the employee feels that that function was not his responsibility

The last aspect is more common than is realised in the pig industry and can frequently lead to a breakdown of the employer-employee relationship.

The job description should include points 1 to 11 in Table 6.2 and that for a farrowing section head might be along the lines indicated.

Table 6.2 Suggested basis of a job description for a farrowing section head

1. Job title	(e.g. stockperson in charge of farrowing section)
2. Responsible to:	(e.g. to Unit Manager and through him to General Pig Production Manager)
3. Major duties:	(e.g. supervise two farrowing house assistants, organise holiday and off-duty periods and regular inspection of stock in the evenings and at weekends, introduce due-to-farrow sows in good time, provide comfortable conditions for sows and piglets, attend to farrowings, minimise stillbirths and livebirth losses, feed sows to maintain good body condition during lactation, ensure sound creep feeding management of piglets, attend to fostering of poor pigs (underweight and sub-optimal body condition) at weaning on to nurse sows, keep medicines in refrigerated store, stock-take and order new supplies in good time, maintain a high standard of hygiene, keep specified detailed records on farrowing, weaning and mortality and summarise weekly, monthly and annual trends promptly, report on farrowing house performance and problems at monthly management meetings, attend to training, motivation and evaluation of farrowing house assistants and assist with their job ranking and grading)
4. Training	(e.g. attend: standard 2-day company induction course, 3 days (or equivalent) on-job training per year, 4 half days per year within-company advisory discussion days, 4 evening pig discussion group meetings per year, occasional pig conferences and pig industry fairs)
5. Occasional duties	(e.g. relief duties in 1st and 2nd stage nursery to cover for illness and other contingencies)
6. Food supplies	(e.g. assess usage and supplies of lactation diet and creep feed weekly; give one week's notice of need for new supplies)
7. Equipment	(e.g. attend to routine maintenance of equipment such as feed barrows and power washer. Prepare monthly inventory of materials and equipment (thermometers, light bulbs, heat lamps, bedding) and order supplies, review need for repairs and maintenance; submit written proposals monthly to unit manager)
8. Necessary skills and requirements	(e.g. recognise signs of imminent farrowing and abnormalities during farrowing, skilled in assisting farrowing sows and newborn piglets as required; recognise mummified piglets, stillbirths and early post natal deaths to ensure accurate

64

recording; skilled in intensive care of underprivileged newborn piglets and in fostering so as to maximise wellbeing and survival; recognise early signs of ill health in sow and piglet and administer prompt remedial treatment; meticulous and honest in record keeping; develop good team spirit In farrowing house staff; be considerate and patient with animals and never mistreat)

9. Working environment

(e.g. warm showers provided – showering in and out obligatory (separate facilities for male and female workers); clean overalls, light caps and wellington boots provided; lock up cabinets for personal clothing and valuables; rest room area has heating, an electric kettle, microwave oven, refrigerator, crockery and cutlery, washing facilities, simple leisure pursuits, e.g. card games, dominoes and draughts, adjacent toilets and payphone; small office has desk, filing cabinet, refrigerator for medicines, pinboards on wall for summaries of farrowing house records and graphs indicating performance trends). Farrowing house temperature maintained at 20°C with provision for required higher temperature and comfort levels for piglets in their creep areas; adequate natural and artificial lighting at all times; good air flow (fresh environment) and low dust levels

10. Working conditions

(e.g. alternate 40 and 48 hour working weeks. Weekday working hours 07.00 to 16.00 hrs with 3 rest periods per day totalling 1 hour to be used at discretion of stockperson in charge in consultation with team-mates but usually allocated as AM (15 minutes), lunch-time (30 minutes) and PM (15 minutes). Four hour daily work period on Saturdays and Sundays to be organised by stockperson in charge and team-mates so as to best fit in with needs of stock and the staff. Two weeks paid holiday per year plus other statutory holidays)

11. Job ranking

(e.g. this will be based on skill levels, training evaulation, ability to solve problems, leadership ability, responsibilities and length of service. The farm operates a system of grading stockpeople on a 1 (new inexperienced recruit) to 5 (senior, highly skilled stockperson) basis. Grading is based on a combination of internal and external (training provider) assessments and both monetary and non-monetary rewards are in accordance with the job grading attained)

As a basis for effective advertising and recruitment a 'person specification' must also be prepared. This should define the background education, training and other characteristics of the type of person best suited to fill the vacancy.

SCOPE FOR JOB FLEXIBILITY

With the impending serious decline in the number of teenagers available for employment, the pig industry must (as some other industries have already done) adopt more flexible arrangements for employment. For example, part-time employment and job sharing should be encouraged. Married women rearing young families may not be available for full-time employment but 2 such people might be able to job-share, each working a half-day and thus covering a whole working day between them. This can lead to effective recruitment of these women as full-time employees when their children grow up.

WAGES AND PROMOTION PROSPECTS

High initial wages may attract applicants to a pig stockperson post for the wrong reasons, i.e. they may be more influenced by the attractiveness of the wage than by that of the job. It is interesting in the study of Howard and his associates in Canada that the larger farms which attracted more applicants for available stockperson posts did not necessarily have higher wages but they did have greater prospects for promotion. Thus, a career structure affording good promotion prospects within a farm would appear to be more attractive to the appropriate type of employee than high initial wages. However, initial wages offered must not be too far adrift of industries competing for the available labour supply. Curnow (1989) reported the results of a survey of personnel managers in industry other than agriculture. Of 1000 such managers interviewed, 66% indicated that they had increased their basic rate of pay during the preceding 2 years in order to improve recruitment.

Selecting employees

Having attracted a good number of potentially suitable employees, the next step is to select the most appropriate one for the post on offer. There are several important steps in this process.

WRITTEN APPLICATION

Cleary (1990) advocates that a written application be sought from each applicant on a form provided by the employer. He stresses the likelihood of overlooking important information if selection is based solely on a casual personal interview. The recommended range of information sought in an application form is summarised in Table 6.3.

Cleary has found that this type of information serves as a useful initial 'screening device prior to interview and can enhance the scope of the interview for both the employer and the applicant.'

Table 6.3 Suggested format for a job application form

- Name and personal details
- Standard of education/qualifications/training achieved
- Details of previous employment (last 5–7 years) including positions held and reason for leaving
- Details of pig production knowledge/skills/experience
- Details of current and past health status
- Names of two referees
- Other information determined by the employer

Source: Cleary (1990)

INTERVIEW

The selection interview should be used to both complement and supplement the information supplied in the candidate's written application and the standard completed application form.

'When you select the person, you select a potential contribution to profit' (Freemantle 1985). This phrase summarises quite well the attitude that must be taken when interviewing potential staff. The quality and dedication put into this process will be creating an image of the company in the mind of the interviewee. A blunt, hurried and careless interview will give the image that 'people are not too important', and that 'carelessness is accepted in this company'. Sticking to the exact date and time arranged are good examples of attention to detail.

Two things must be accomplished in this interview: first, proving that the candidate is up to the company's expectations, and second, persuading the candidate that the company is worth joining. A bad initial impression of the company might create the subconscious feeling in the mind of the candidate that this will be a temporary job.

A basic brochure (for handing out) describing the company's activities, size and performance and some policy aspects can be useful to create a positive company image.

The written information supplied can be cross-checked against the details of the job specification to guide the interview towards supplementing the facts already obtained. The applicant should be put at ease so that he/she behaves as naturally as possible. Such a relaxed atmosphere should be conducive to the applicant conveying to the interviewer impressions of attitudes and aspirations.

The arrangements for and atmosphere prevailing during interviews of candidates for a stockperson post are most important. 'When you select the person, you select a potential contribution to profit.' A badly prepared and hurriedly conducted interview in sloppy surroundings will give the impression that 'people are not too important' and that 'carelessness is accepted in this company'.

SELECTION PROCEDURE

The best known personal profile plans for staff selection are the Rodger '7 point' system and the Fraser plan. The system proposed in this book will use these two plans as a baseline to produce a simplified version adapted to pig production. It is not the purpose of the proposed plan to produce a psychological profile of the interviewed person. The system proposed here must be seen as an aid to those pig farm managers who select staff without the use of professional recruitment services.

The following are the aspects to be evaluated in the selection procedure:

1. Physical attributes—Pig stockmanship must be considered as a job with relatively high physical demands; good general health and fitness are desirable features in a candidate. Such health problems as hay fever or respiratory diseases, deafness, poor eyesight even with glasses, chronic diseases and serious physical handicaps are usually likely to affect performance. Cleary (1990) emphasises the importance of a medical examination, which should include eyesight and hearing tests because these functions are important to the perceptual skills of the stockperson. Cleary points out that such examinations provide a baseline for both parties with regard to any subsequent compensation claims by employees.
2. Mental attributes—Pig stockmanship as a job does not necessarily demand special mental attributes such as creative or numerical ability. However, if these abilities are present they will prove helpful to the individual in relation to promotion and eventual attainment of a managerial position, if the stockman has such aspirations.
3. Education and qualifications—Training to diploma, degree or certificate level may be useful for section or unit managers, but over-qualified employees may become easily dissatisfied with their job if they are allocated to 'general stockman' duties. When studying the relationship between educational background and performance, Segundo (1989) found no significant difference in stockmanship between groups with different qualifications. However, stockpeople with diplomas, degrees and certificates did have higher average stockmanship scores (although not significantly higher) than the others interviewed.
4. Personality—Personality problems all too often prevent well qualified and skilled employees from producing good results. Key factors are motivation and ease of social adjustment (e.g. to the farm's existing team of workers). Particular personality characteristics that are important in pig stockpeople working on large farms are patience, ability to pay attention to detail, perseverance,

being easy going, considerate when handling animals and at least moderately sociable.

5. Special circumstances—Some jobs make special demands over and above those listed above. For example, in attending farrowings, assessing the need for helping the sow and in helping to establish newly born piglets, particularly the small and otherwise disadvantaged ones in large litters, a high degree of 'maternal instinct' and empathy with the sow and underprivileged piglets is required. This need places an inevitable bias towards female stockpeople in this area, although some men are also well endowed with strong 'maternal' instincts in looking after the sow and her litter and do an excellent job as farrowing house attendants.

INCREASING THE OBJECTIVITY OF THE CHOICE

The most effective way to try to maximise the chances that the right candidate is selected for the job is to base this on a combination of the following procedures:

- a study of the written application
- scoring each candidate on the basis of criteria contained in the application form and obtained at interview
- subjecting a short list of candidates to a trial period on the farm
- pooling all information at a meeting of management and staff to facilitate final decision making by management

A possible basis for these procedures will now be outlined.

Objective scoring of candidates on the basis of the completed application form and the interview

A great deal of information and impressions will have been obtained from each candidate and each will have their own particular strengths and weaknesses. It is extremely useful to attempt to give a score to each piece of relevant information and then to calculate an overall score for each candidate.

The basis of such a scoring system is suggested in Table 6.4. This system has not yet been tried and tested but the principles inherent in it should be considered by employers who can amend it as required to suit their own particular objectives.

Subjecting a short list of candidates to a trial period on the farm

Using a trial period as part of the selection procedure is particularly suitable for new recruits to the pig industry and has 3 main advantages: (1) It allows for assessment of aspects that cannot be evaluated in an interview, such as social adaptation and motivation. (2) It allows each 'trialist' the opportunity to find out if they really like the job and the work environment before committing themselves to it on a permanent basis. (3) It allows expression of opinion of the other members of staff. This third point is important for the following reasons:

- it takes into account the existing group or team structure
- the involvement of staff in such a process is not only a demonstration of respect towards the team, but also gives them some control over one of the factors which, according to Mayo (1930), has the largest influence on the staff's attitudes towards work i.e. the group, its togetherness and team spirit
- responsibility (accorded to staff) was considered by Herzberg (1959) to be one of the strongest motivating factors
- it relieves, at least in part, the responsibility of management for selecting individuals who might eventually create friction within the staff

The new employee will be allocated work 'side by side' with experienced stockpeople on a rotational basis (say a week in each section of the unit). This will allow him or her to become familiar with the different tasks and people in the piggery. This will also allow the stockpeople who work with him/her to get a feel of the trialist's abilities, skills and attitudes.

The length of the trial period can be variable according to each particular situation, but it should probably be not less than 2 weeks and no longer than a month. At the end of the trial period each stockperson with whom the trialist has worked should be asked to complete an assessment form along the lines of that shown in Table 6.5.

Final decision making
The final decision regarding the suitability of the new 'employee' for the job should follow a staff/management meeting in which all relevant information from the application form, the interview, the scores allocated on the basis of Table 6.4 and following a trial period (Table 6.5 if such has taken place) is considered thoroughly and objectively.

Table 6.4 Illustrative example for the rank scoring approach to selecting a pig stockperson from job applicants for a post in the farrowing section

Assessment Sheet	Scoring Sheet
NAME .... DATE	
A. *Information derived from application form*	
1. Age	
Under 18 years old ☐	−2
18–24 years old ☐	0
25 years old and over ☐	+2
2. Sex[A]	
Male ☐	0
Female ☐	+2
3. Marital Status	
Single ☐	0
Married ☐	+2
4. Education	
No school-leaving qualifications ☐	−2
School-leaving qualifications ☐	0
Further education courses completed ☐	+2
Vocational training completed ☐	+2
5. Health[B]	
Hearing ☐	
Eyesight ☐	
Hay fever ☐	
Respiratory diseases ☐	
Other health problems ☐	
6. Interest in pigs before starting this job	
Highest rating ☐	+2
Next highest rating ☐	+1
Average ☐	0
Below average ☐	−1
Lowest rating (nil or negligible interest) ☐	−2
B. *Information derived from the interview*	
7. Main reason for applying for the job	
Wants to work with pigs ☐	+2
Wants to work in agriculture ☐	+1
Wants to live in the country ☐	0
Other ☐	−1

8. What do you think is the single most important attribute of good stockmanship

Being patient yet persevering with animals	☐	+2
Paying attention to detail	☐	+1
Being able to get through the work as quickly as possible	☐	−1
Being able to force animals to do what you want them to	☐	−2

9. Which of the following statements suit you best (pick any two of the six options)

I consider it very important to have a good working relationship with my fellow workers	☐	+2
I prefer team work although I don't mind working alone	☐	+1
I prefer to work on my own^C	☐	−1
I will enjoy taking greater responsibility once I know my job	☐	+1
I am very keen to attain higher vocational qualifications within the pig unit as quickly as possible	☐	+2
I am very keen to get promotion as quickly as possible^D	☐	−2

10. Attitude to training
How important is the provision of job training to you

Very important	☐	+2
Moderately important	☐	+1
Neither important nor unimportant	☐	0
Moderately unimportant	☐	−1
Largely unimportant	☐	−2

11. Do you have other skills that you consider could be useful to this company (Interviewer notes these below)

_____

_____

Total suitability score   E

(A) Scoring is for farrowing and first stage post-weaning nursery jobs, for which a female is given a higher score. For all other jobs a zero score would be given to both male and female applicants.
(B) Only negative scores given depending on the severity of the disfunction and according to its relevance for operation of the system. If the applicant is healthy, a score of zero is suggested.
(C) Negative score would not apply in a single man unit.
(D) This negative score is relative to the promotion prospects that the particular company offers: most pig farms offer poor promotion prospects and people with strong promotion ambitions will frequently become dissatisfied with the job.
(E) The maximum score is +24 while the minimum score is approximately minus 11 (this score can be even more negative if the applicant has any health problems as specified in section 5). It is suggested that a minimal suitability score for acceptance in the farrowing section is +10. Such a scoring system helps to balance considerations of the strengths and weaknesses of the various candidates and will be extremely useful in helping to decide on the final choice.

Table 6.5 Suggested form for evaluating a prospective new employee following a trial period on the farm, to be filled in by an established stockman (mentor/trainer) and the manager's score sheet (with an example assessment shown)

SENIOR STOCKMAN'S (TRAINER'S) FORM	Score sheet used by team leader/ manager	

In strict confidence

Name of mentor: ....
Section of farm: ....
Time with new recruit: .... (number of days)

	Scoring system used[A]	Score given to an applicant[B]
A. Social adaptability *Tick one box*		
(of new recruit in relation to present employees)		
I consider the new recruit to be:		
• an excellent person to work with ☐	+2	
• an easy person to work with ☑	(+1)	+1
• a somewhat difficult person to work with ☐	−1	
• a most difficult person to work with ☐	−2	
B. Workplace adaptability		
(of new recruit in relation to production system)		
It seems to me that the new recruit		
• has settled in very well ☐	+2	
• has settled in reasonably well ☑	(+1)	+1
• is still unsettled but is likely to settle in soon ☐	−1	
• is still unsettled and is unlikely to settle in ☐	−2	
C. Animal care		
(shown by new recruit in relation to animals in unit)		
In my judgement the new recruit has:		
• shown an excellent standard of care ☑	(+2)	+2
• shown a satisfactory standard of care ☐	+1	
• tended to be neglectful of animals ☐	−1	
• been very neglectful of animals ☐	−2	
D. Work attitude		
(of new recruit in the production unit)		
I consider the new recruit's attitude to be:		
• highly responsible ☑	(+2)	+2
• satisfactory ☐	+1	
• unsatisfactory ☐	−1	
• not at all responsible ☐	−2	
E. Overall assessment		
(of new recruit as a stockperson)		
Taking everything into consideration, how would you assess his/her future?		
• will cope very well ☐	+2	
• will cope reasonably well ☑	(+1)	+1
• will have difficulty coping ☐	−1	
• will not cope at all ☐	−2	

Other comments

What other observations can you offer based on your experience of working with the new recruit?   *Yes No*

• Does he/she learn quickly ☐ ☐		*Total score*
• Is he/she patient with the animals ☐ ☐		*+7*
• Does he/she pay attention to detail ☐ ☐		
• Is he/she tidy and methodical at work ☐ ☐		
• Is he/she inclined to lose his/her temper ☐ ☐		

(A) The maximum and minimum scores possible in this assessment are +10 and −10 respectively. It is suggested that a *minimal acceptable score*, averaged over all the stockpeople making the assessment, is +5. (B) In the example given, the applicant has come out with a score of +7, which makes him/her an acceptable recruit.

Summary

This chapter has dealt with the most important objective of attracting good applicants for a stockperson post and then selecting the best one for the job. The main considerations involved in these 2 processes are summarised in Figures 6.4 and 6.5.

Having made the best choice from the applicants for the post, the next step is to ensure that he/she is properly inducted into the new job, integrated into the existing team and given every support thereafter in the form of continuing education and training, motivation and good working conditions. Such provisions will help to ensure that the new employee becomes an increasingly effective member of the working team and is retained in employment for a long period of time. These latter aspects of inducting and retaining new employees form the basis of the next chapter.

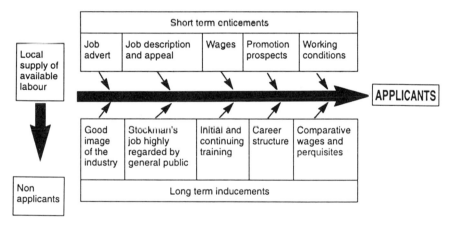

Figure 6.4   Mechanisms to attract applicants to a stockperson post

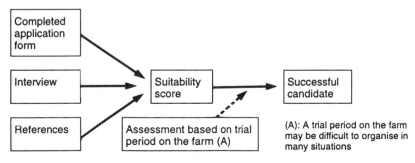

Figure 6.5   Basis of selecting the best applicant for the job

7

Inducting and retaining new employees

The attitudes of staff towards the company will be a reflection of management attitudes towards staff.

The only new stockperson who is hired more or less at zero cost is the son or daughter of the owner. However, as pointed out previously, employees from the owner's family are becoming less common in the pig industry, and most new entrant stockmen come at a significant cost. Few owner/managers take time to add up how much they really invest in the recruitment and selection of their new employee. In this process, time and money will have to be spent on several or all of the following:

- drafting the job description
- placing the advertisement
- preparing the job application form
- handling enquiries and distributing application forms
- sorting out the applications and selecting a short list of candidates
- arranging interviews
- holding interviews
- deciding on the best candidate and sending out the job offer

It will be the exceptional stockperson who gives much thought to such costs. Yet they are real costs for the industry and influence its competitiveness. For this reason, the investment does matter to the employee/stockperson as well as to the employer/manager.

Failure to retain a stockman incurs a further cost to the business unit. Employee turnover can be a 'disruptive, costly process' according to Howard and his colleagues, based on their studies of labour on pig farms in Canada. The first step in getting the best out of a new employee of good potential is to ensure good induction procedures.

Induction of new employees

Induction is the process of introducing recruits to an organisation and explaining their role within it. McCreadie and Phelon (1947) found that the way new employees are 'introduced to the job' was an important reason for labour turnover. The impressions gained by new employees during this period can influence their perception of the firm for many years to come. Also, good induction procedures help employees to fit into strange and initially uncomfortable environments. The basic objectives of induction of new employees are summarised in Table 7.1.

Address of welcome (Bienvenidos) to a large pig farm in Mexico. The welcome and careful guidance of new recruits to the pig unit staff are essential features of good induction procedures.

Cleary (1990) from his experience of pig production enterprises in Australia also emphasises the crucial importance of effective induction of new stockpeople in the following terms: 'Poor induction of new piggery workers can lead to discontent, low productivity, poor quality work and high labour turnover.' He infers that such carelessness in introducing new workers plays a significant role in the fact

77

Table 7.1 Summary of the basic objectives of good induction procedures for new employees

Induction is the process of familiarising a new recruit with:

- the organisation of the business unit
- the work tasks to be undertaken
- the knowledge to be acquired
- the skills to be acquired
- the production targets for the unit and the role of a stockman in helping to attain these
- the animal welfare codes of the unit and the role of the stockman in achieving the required standards
- the methods used for communication of information within the stockpeople/manager team and the contribution expected of the new stockman
- the training policy of the business unit

that labour turnover rates are highest among stockpeople with only a few months' service. Cleary found that 'proper induction procedures help workers to adjust quickly to new jobs, stimulate their interest, win their loyalty and assist them to make an effective contribution.' In view of the scarcity of potential recruits to the pig industry it is vital to give carefully selected new employees on pig farms every possible incentive to stay in employment and appropriate guidance and training to become increasingly more competent stockpersons. Much more time is wasted in seeking and selecting new employees than is devoted to proper induction and training procedures according to Cleary, whose suggested basis of good induction procedures is outlined in Table 7.2.

Newcomers usually join the pig unit wanting to do a good job and to be accepted by colleagues and generally to become a part of the organisation. Good induction procedures should help recruits achieve these objectives. Recruits should be informed or be encouraged to ask what to do if he or she:

- has a problem in understanding the wages system
- has a medical problem
- feels that working conditions are unsafe
- does not get on with other people in the department
- has difficulty with the work
- feels harassed or bullied
- does not receive adequate training
- has any other complaint

Table 7.2 Outline of good induction procedures of new employees to pig farms, as suggested by Cleary (1990)

WORKER'S NAME.. DATE ....
TITLE .... DATE REPORTED FOR WORK ....
ORGANISATION SECTION ....

Purpose. This list provides an outline to follow in welcoming and systematically inducting new workers. The induction period is the greatest opportunity to win the workers' loyalty, stimulate their interest, and get them to work effectively. Check each item to be sure you have welcomed the new worker properly and have provided all the information that he should have.

Date completed

1. Get ready to receive the new worker
(   ) Review work experience, education and training   ....
(   ) Have an up-to-date description of the job or a list of duties
       and responsibilities available for discussion   ....
(   ) Have the workplace, tools, equipment and supplies ready.
       This makes the wroker feel that he has come to a real job   ....

2. Welcome the new worker
(   ) Greet him personally, put at ease, make welcome, show
       friendliness by using name often   ....
(   ) Indicate your relationship to the new worker   ....
(   ) Assign a workplace, tools, equipment   ....

3. Show genuine interest in the worker
(   ) Discuss background and interests   ....
(   ) Enquire about housing situation   ....
(   ) Enquire about transportation to and from work   ....
(   ) Enquire about any possible financial difficulties because of
       the pay lag and suggest local sources of assistance   ....
(   ) Provide learning aids such as industry journals, manuals, job
       instruction procedures, list of special or technical terms   ....
(   ) Explain use and care of tools and equipment   ....
(   ) Stress safe working habits   ....

4. Follow-up
(   ) Maintain regular contact with the new worker. Assist in
       developing a sense of belonging by enquiring about any
       difficulties   ....
(   ) Encourage questions but do not interrogate   ....
(   ) Be alert for personal problems or personality clashes that
       could affect the newcomer's performance   ....
(   ) Listen carefully to any expressed dissatisfaction   ....
(   ) Correct errors with further instruction, not criticism. Do not
       reprimand in public as grievances may set in.   ....
(   ) If performance improves, be generous with comments.
       Express them with sincerity   ....
(   ) If performance warrants, reward appropriately   ....

Induction procedures do not alter a worker's character or personality, but they usually influence behaviour and performance. If induction procedures enable the new worker to gain satisfaction, responsibility and increased skills, the management task of the piggery is lightened.

Some of the problems with induction are management's lack of time set aside for transmitting relevant information and the unsuitability of the environments in which inductions sometimes take place. No new recruit is capable of absorbing large amounts of information all at once (e.g. on the range of fostering procedures to meet widely varying needs). Therefore, induction should be gradual and spread over time.

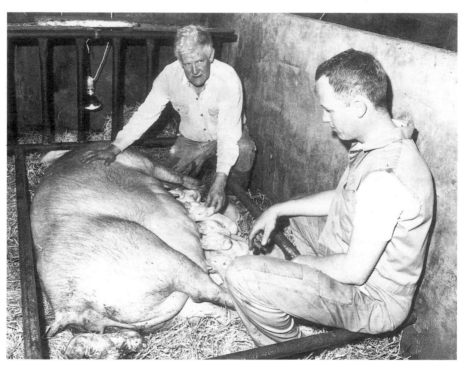

Proper induction procedures help workers to adjust quickly to new jobs. They also help to stimulate their interest, win their loyalty and assist them to make an effective contribution. Here an experienced stockman guides a new recruit on assistance measures for newborn piglets and on the needs and procedures for fostering.

The induction usually begins with a guided tour of the buildings, in which the employee is introduced to the people with whom he or she is going to work, and an effort must be made to try to make the recruit feel welcome. After this, it is advisable to have the new recruit in a private office where further explanations will be given. Organisational structure, promotion and training opportunities, as well as the recruit's duties, responsibilities and expected performance must be explained in full at this stage. Other important points to be explained include how to report absences, overtime commitments, location and operation of fire fighting equipment, first aid facilities and provision of protective clothing. Other smaller matters such as provision and arrangements for private phone calls, meal breaks and hygiene requirements must be explained. Segundo (1989) found that some of these 'smaller matters' such as the unavailability of the telephone for urgent but infrequent private calls during working hours are capable of breeding great dissatisfaction in employees.

Often, new jobs are associated with new lifestyles, unfamiliar locations and work routines, as well as new relationships, and thus might create high levels of anxiety. Recruits can easily feel bewildered and unwanted by the existing staff, so sympathy and understanding are needed during this acclimatisation period.

Cleary concludes that 'good induction procedures should be second nature to a good piggery owner/manager and need not tie up too much time if done little by little on a regular basis'.

Retention of employees

According to Howard and his associates in Canada, although culling poor performers and social misfits can be beneficial, 'employee turnover is often a disruptive costly process which most employers want to minimise.' Cleary in Australia echoes this view. 'High staff turnovers have been a feature of the industry in the past and the loss of trained staff continues to disrupt piggery operations and to hurt productivity and profitability.' Worse still, staff who leave a pig farm are often also leaving the pig industry. Beynon (1990) in England states 'an increasing number of our skilled personnel are leaving both pigs and agriculture for better prospects, pay and conditions in other industries.'

Staff turnover on pig farms is therefore a critical issue.

LEVEL OF TURNOVER OF PIG STOCKPEOPLE

In the study of Howard and associates on pig farms in Ontario, Canada, the average staff turnover was every 5.7 years (range 0.35 to 42 years). This was lower than that in non-agricultural industries.

The labour turnover rates in manufacturing, finance and health care in 1978–80 were 4.6, 3.5 and 3.6 years respectively while that among staff of the Ontario Ministry of Agriculture and Food was estimated in 1989 at 4.2 years. Goodman (1990) cites a survey of 25 pig farms in East Anglia, England, in the late 1980s. Of a total of 83 employees, 47% had been in the current job for 5 years or less. Segundo and associates, in a study on 15 farms in Scotland in 1989, found that the 65 stockpeople interviewed had been in their current job for an average of 4.1 years. Cleary's (1990) assessment of pig farms in Australia was that high staff turnover was a feature of the pig industry in recent years. Thus, a high turnover rate among pig stockpeople appears to be a problem in many countries of the world.

REASON FOR HIGH TURNOVER RATE AMONG PIG STOCKPEOPLE

In the study of stockpeople in Ontario, Canada, Howard and his colleagues found that 55% of stockpeople planned to leave their current employer within 3 years, 25% wanted to own their own farm, 21% sought non-farm work while 9% expected to change to another pig farm within 2 years. Those who sought non-farm work did so with the objectives of achieving higher wages and more time off as well as for expectations of feelings of greater achievement and recognition in their new job. When stockperson turnover was related to farm and employer characteristics, the relationships summarised in Table 7.3 were found.

Poorer wages, benefits, job satisfaction and farm profits appeared to predispose to higher turnover rate of stockpeople. The older the boss and the greater the number of hours worked per week the greater was the staff turnover. However, older workers were less likely to leave their job. Managers having the belief that stockpeople need to be motivated more to get the best out of them (Theory X Type) supervised workers more closely and provided incentives to encourage greater output and efficiency, and these appeared to have lower turnover among their staff. Theory Y managers, who had a higher turnover rate among their staff, believed that stockpeople will make a big effort because of the basic inherent motivation in people to perform well. Therefore these latter managers placed more responsibility on staff, did not supervise them so closely but expected stockpeople to achieve high output and efficiency because of the greater job satisfaction derived from such high performance.

Table 7.3   Factors affecting staff turnover on pig farms in Canada

	Factors having a strong influence	Factors having a slight influence
Reasons why stockpersons left pig farms	• Low cash wage • Little job satisfaction • Management style which provides little motivation to workers	• Few or no fringe benefits • Lack of extra payments • Low farm profits • No other employees • Job advertisement
Reasons why stockpersons stayed on pig farms	• Reasonable number of days in working week ensuring a reasonable wage • Employer in older age group	• Reasonable number of hours in working day ensuring a reasonable wage • Sufficient holidays • High number of sows per farm thus ensuring a reasonable number of workers • High number of pigs per sow • Payments-by-result scheme in use

In addition to the aspects examined by these Canadian workers, other factors such as the provision or non-provision of continuing vocational training can affect the rate of turnover on pig farms

Source: Based on Howard and associates (1990)

An intriguing outcome of the study in Ontario was that despite pig stockpeople working longer hours, receiving lower wages and having fewer benefits than workers in non-agricultural industries, their turnover was lower than among those working outside agriculture. This may indicate that while wages are important, non-economic factors constitute a large part of job satisfaction. Howard and associates contrast jobs as pig stockpeople with high-tech computer jobs where the wages are high but so also is employee turnover.

Factors affecting job satisfaction on pig farms

The work of Segundo and associates in Scotland on job satisfaction which was conducted on 65 stockpeople from 15 pig farms is worthy of detailed consideration. The number of breeding sows per farm ranged from 180 to 1750, while the number of employees per farm ranged from 1 to 13. Using a list of 56 job related factors, stockpeople were asked to express the degree to which they were satisfied or dissatisfied with these in their present job. The proportion of interviewees dissatisfied (extremely, very and moderately) was used to rank the factors from the highest to the lowest degree of dissatisfaction. Table 7.4 presents the 25 factors with which interviewees were most dissatisfied.

Almost half (49.2%) of the respondents were dissatisfied because of the lack of bonus or incentive schemes, while a quarter (24.7%) were dissatisfied with their basic wage. A generally high level of dissatisfaction was expressed at the failure to hold or make adequate provision for (1) 'formal meetings' (44.6%), (2) discussions on the pig enterprise in terms of physical performance (32.3%), (3) discussions on technical matters pertaining to pigs (30.8%) and (4) discussions on financial aspects of the enterprise (26.2%). More than a quarter of stockpeople (26.2%) were dissatisfied because of the lack of recognition of their achievements by the owners/managers. Approximately one third of the stockpeople were dissatisfied because of the absence or insufficiency of arrangements for on-farm and off-farm training (35.4%, 33.9% and 33.9% of stockpeople were dissatisfied because of the poor provisions made in terms of factors 3, 4, and 5 respectively). A high proportion of stockpeople were dissatisfied with the communication between management and themselves and with their lack of involvement in decision making (the proportion of stockpeople dissatisfied with factors 2, 7 and 22 were 44.6%, 32.3% and 21.6% respectively).

Concern about aspects of pig wellbeing and performance was evident by the proportion of stockpeople who were dissatisfied with the pigs' health (29.3%), pig housing and equipment (30.8%), the general comfort of the pigs (27.7%), with the general environment, particularly dust levels (32.3%), the slow rate of introduction of new technology (27.7%), the difficulty of recruiting good staff (20.1%) and with the level of pig performance being achieved on their farms (26.2%). With reference to kitchen facilities for stockpeople (factor 21), it can be seen that approximately half (53.3%) of the farms did not have these facilities (mainly the small farms) and approximately

Table 7.4   The factors with which stockpeople were most dissatisfied

Factors	% of stockpeople dissatisfied	% of farms which did not have the factor[A]
1. Incentive/bonus schemes	49.2	100
2. Formal meetings of management with staff to discuss farm performance	44.6	93.3
3. Opportunities for further training	35.4	
4. The provision of on-farm training	33.9	86.6
5. The provision of off-farm training	33.9	53.3
6. Involvement in decision making	32.3	
7. Communication from management to staff	32.3	
8. Discussions between management and staff on farm physical performance	32.3	33.3
9. Quality of environment in pig buildings	32.3	
10. Quality and maintenance of pig housing and equipment	30.8	
11. Discussion on technical aspects	30.8	46.6
12. Health of pigs	29.3	
13. Introduction of new technology	27.7	
14. Comfort of pigs	27.7	
15. Relationship with fellow workers	27.7	
16. Discussions on financial aspects	26.2	93.3
17. Recognition of achievements	26.2	
18. Farm physical performance	26.2	
19. Financial reward (basic wage)	24.7	
20. Career prospects	24.7	
21. Kitchen facilities	21.6	53.3
22. Opportunities for communication from staff to management	21.6	
23. Informal meetings with staff	21.6	33.3
24. Provision of protective clothing	20.1	6.6
25. Good staff recruitment	20.1	

(A) Some of the factors considered were not operated or present on all farms. The degree of dissatisfaction with factors may be related to either the absence of the factor, or to the adequacy of its provision.

a fifth (21.6%) of all stockpeople expressed dissatisfaction with the lack or quality of this provision. Approximately a quarter (24.7%) of stockpeople were dissatisfied with their career prospects, and/or their prospects of promotion (factor 20). Other aspects with which stockpeople were dissatisfied included their relations with fellow workers (27.7%) and the lack of provision of adequate protective clothing (20.1%).

Another method was used to assess the factors contributing most to job dissatisfaction. An open question was asked at the end of the questionnaire: namely 'What is the most annoying or dissatisfying aspect of your job?' The replies were grouped into 9 categories and are presented in Table 7.5.

Table 7.5 Main aspects of dissatisfaction among stockpersons

Sources of dissatisfaction	Number of dissatisfied stockpeople	Dissatisfied stockpeople as a proportion of all stockpeople (%)
Attitudes of other workers	16	23.9
Aspects of the management style	16	23.9
The work itself	8	11.9
The physical environment on the farm (mainly dust levels)	7	10.4
The performance of the pigs	4	6.0
The health and comfort of the pigs	4	6.0
Pay and conditions	3	4.5
Lack of prospects of promotion	3	4.5
Other aspects (e.g. aspects of company policy)	6	8.9
	67*	100.0%

* Two stockpeople could not decide which of two factors annoyed them most, so both factors were taken into consideration

The attitudes of other workers which caused annoyance and lowered the job satisfaction of their fellow workers included careless-ness, lack of willingness to cooperate with others, and lack of commitment to the work which had to be done. Aspects of management style which reduced job satisfaction included poor communication between management and staff, lack of involvement of stockpeople in decision making, and failure to recognise the achievements of stockpeople. Aspects of 'the work itself' which caused dissatisfaction included power washing and pen cleaning after stock were removed from the pen or the house, and excessive hours of weekend working. The dissatisfaction with the physical environment on the farm applied mainly to dust inside the pig buildings. Aspects of 'performance of the pigs' which caused dissatisfaction among stockpeople included such factors as low number of piglets born alive, high piglet mortality, and low and/or 'inexplicable fluctuations' in conception rate. Complaints about 'health and comfort of pigs' which caused dissatisfaction included mortality due to pneumonia, housing related diseases and rough or cruel handling of animals by other workers. Factors which annoyed under the heading of 'pay and conditions' included the low basic wage and insufficient holiday and/or over-time pay.

When added together, the aspects included under the headings of 'attitudes of other workers' (23.9%), 'aspects of management style' (23.9%) and 'lack of prospects for promotion' (4.5%), it can be seen that in 52.3% of cases, the most dissatisfying aspects of stockpeople's jobs were related to human behaviour and human resources management. When adding together the 'performance of the pigs' (6.0%), the 'health and comfort of pigs' (6.0%) and the physical environment on the farm (10.4%), it can be seen that 22.4% of the factors which annoyed stockpeople most were concerned with the conditions prevailing on the farm.

The main conclusions from the study of Segundo and his associates were:

1. A variety of factors affected the job satisfaction and hence the labour stability in the farms which participated in the study.
2. A high proportion of the problems were related to the management of the human resources on these farms. Considerably more attention should therefore be devoted to this aspect by owners and managers of pig farms and by agricultural, educational and advisory organisations.

Thus, in response to the question on the most dissatisfying aspects of the job, pay and conditions along with prospects of promotion came fairly low on the list. Aspects of management style, the attitudes of other workers, some components of the work itself (repetitive mundane jobs such as pen cleaning and power washing, excessive hours of weekend working and dust levels in the pig houses) and provisions for the general health and comfort of the pigs were the factors which concerned stockpeople most.

In view of the generally high turnover of labour in the pig industry and the indications that current wage level is far from being the main bone of contention in most situations, it is important to look closely at non-pay considerations and how these can be improved for pig stockpeople.

It is likely that several important changes have occurred in the pig industry over recent years almost unnoticed, perhaps because they have taken place gradually, and these may be at the core of a high degree of dissatisfaction among stockpeople which has in turn contributed to high turnover of staff. Such changes include the increase in size of pig farm and in the number of workers employed. There has also been a change in the source of supply of workers with fewer coming from the traditional sources of the families of farmers and farm workers and an increasing proportion from outside agriculture, often with little or no experience of agriculture or farm animals. Because of the increase in the size of farms, the operators of previously small units, who had to be skilled mainly in pig management, have now to become increasingly competent in 'people' management. With the increase in the proportion of inexperienced stockpeople entering the industry the need for training is highlighted, while the increasing number of workers in a large pig unit needs managers who are good with people and can mould them into an efficient working team.

It is highly probable that many of the sources of dissatisfaction, detected in stockpeople by Segundo and his associates in their study in Scotland, reflect the above changes in the industry and the failure, as yet, to adapt.

Thus, there was dissatisfaction with the amount of training received. Aspects of management style also caused annoyance including poor communication between management and staff, lack of involvement of stockpeople in decision making and lack of recognition of achievements. Some stockpeople were upset by the carelessness, lack of commitment and lack of willingness to cooperate on the part of some of their colleagues. Thus, they were not an effective working team and the team spirit was poor. There were insufficient discussions about herd performance and about the general policy and operation of the unit. Thus, many workers felt a lack of involvement. They wanted to be part of a team effort in striving towards the objectives of the farm but they were being denied that privilege. Many vainly sought more influence on and responsibility for decision making. Of course, for staff to gain more responsibility for decision making it is necessary for management to be confident that staff have a high capability of making the correct decisions. The likelihood of making the correct decisions, in turn, is dependent on knowledge and experience and knowledge reflects on the adequacy of training. So it is likely that some stockpeople saw effective training as one of the main routes towards their gaining more responsibility and influence on the operation of the farm but since they were being denied effective training they were in a 'no win' situation.

A fairly high proportion of stockpeople were dissatisfied with the conditions provided for the pigs (in terms of their comfort and health) and for themselves (lack of kitchen and showering facilities and protective clothing).

Thus there are obvious areas to tackle in order to remove negative influences causing job dissatisfaction and at the same time to increase the provision of positive influences such as better working conditions, training and discussion sessions to upgrade understanding and knowledge and an increase in involvement in decision making in order to build up team spirit and generally to increase the motivation of staff.

The essential first step is to isolate sources of job dissatisfaction and high staff turnover so that the problems can be rectified. After eliminating the factors which have a negative effect on the attitude and motivation of stockpeople it is then possible to start building up the positive influences.

Termination of employment and the exit (termination) interview

Exit interviews are, in a sense, the direct opposite of employment interviews. Resigning subordinates should be encouraged to attend an exit interview (or fill in a form) to discover why they are leaving and hence leave management in a position to implement measures as necessary to prevent others leaving for the same reason.

Some reasons for leaving are unavoidable, such as illness, career development, moving to another area and women who want to devote more time to their children. Examples of frequently avoidable resignations include those caused by personality clashes (often avoidable through lateral staff transfer e.g. to another department) or by dissatisfaction resulting from poor human resources management, or even marginally inadequate pay or working conditions that could be improved to the ultimate mutual benefit of employee and employer. A form that could be used to gather useful information from a departing employee is presented in Table 7.6.

Summary

The main lessons emerging from this chapter are summarised in Figure 7.1.

When potentially good stockpeople are selected for our pig farms,

Table 7.6 Suggested form to be completed when an employee leaves

Date ....

Dear ....,

We regret the fact that you are leaving this company. We hope you are leaving for positive reasons and wish you the very best in the future.

In order to improve our personnel management policy as much as possible, we would be grateful if you could fill in the following form which will help us understand and, if reasonable and possible, correct the circumstances which led you to take the decision to leave this company.

Please tick off your main reason or reasons for leaving:

Thank you for your time.

Yours sincerely,

....

(to be retained by departing stockperson)

		Main reason	Secondary reason
1.	Health problem	☐	☐
2.	Dissatisfied with pay	☐	☐
3.	Poor working conditions (free time, holidays, working hours, etc.)	☐	☐
4.	Poor team spirit among workers	☐	☐
5.	Frictions with members (or one member) of staff	☐	☐
6.	Poor promotion prospects	☐	☐
7.	Attitudes of other workers	☐	☐
8.	Offered better conditions elsewhere (please specify)	☐	☐
9.	Moving to another area	☐	☐
10.	Other reasons (please specify)	☐	☐

(to be handed in to farm office)

Attracting more suitable applicants for pig stockmanship jobs	1. Improve image of pig industry 2. Make the job more attractive 3. Better promotion and advertising of the job 4. Increase job flexibility and opportunities 5. Fair wages 6. Ensure a promotion/advancement ladder

Improving selection of applicants	1. Prepare comprehensive job specification 2. Get each candidate to complete a carefully prepared application form 3. Ensure good interview procedures 4. Score each candidate objectively on the basis of written application form and responses/attitude at interview 5. Consider a trial period on the job for new entrants to the pig industry and arrange for existing staff to grade the candidates 6. Decide on final choice of candidate taking all relevant information and assessment scores into acount

Ensuring effective induction of new appointment	1. Welcome new employee 2. Introduce to future colleagues and work premises 3. Outline duties and responsibilities 4. Indicate to whom he/she is responsible 5. Enquire about adequacy of the employee's housing and financial stability until first wages are due 6. Stress safe working procedures 7. Provide instruction manuals and other informative literature pertaining to the job 8. Encourage worker to bring any problems and uncertainties to the notice of the unit manager

Retaining good employees	1. Provide good working conditions 2. Ensure good continuing education/training opportunities 3. Remove sources of job dissatisfaction 4. Ensure a fair basis for upgrading and promotion 5. Ensure good job satisfaction 6. Provide continuous motivation

Figure 7.1   Attracting and selecting applicants and induction and retention of employees: a summary of the necessary provisions

it is most essential that their inherent strengths and talents are developed by training, motivation and generally by good human resources management.

If one 'golden rule' had to be given for the difficult task of managing staff or stockpeople, it would probably be something like this:

'The attitudes of staff towards the company will be a reflection of management attitudes towards staff.'

If management 'does not have time' for its staff, the staff 'will not have time' for the company. 'Having no time for the company' might mean: lack of interest, lack of concern with the company's objectives, carelessness in handling animals or machinery, many other harmful attitudes and ultimately leaving employment to seek better conditions on another pig farm or, worse still for the pig and the pig farmer, outside the pig industry and agriculture altogether.

Providing effective training for stockpeople, improving working conditions, building up an effective team and generating good team spirit while doing everything possible, within the bounds of available and cost-effective resources, to enhance the motivation of staff are the positive ways ahead and these aspects are dealt with in succeeding chapters.

8

Training

When training is well thought out and executed it is the single most useful tool to improve the performance of stockpeople and probably one of the cheapest approaches to improving the farm performance in general.

Both livestock managers and experienced stockpeople who have worked with and handled farm animals from their early years tend to grossly underestimate their own abilities and skills with livestock. They often take for granted the vast experience and know-how they have accumulated over the years. Hence, many leaders of the livestock industry do not put as high a priority as is justified on the need for adequate training of new recruits and for the continuing education of all stockpeople. This applies as much to pig production as to the other farm livestock systems.

Complexity of stockmanship

When examined in its full perspective, the job of the pig stockperson is a very complex one. Operating a well-designed pig production system can be compared to an expert motor mechanic listening to and testing an engine, so that it may be finely tuned to become more efficient and run more smoothly. However, the effective operation and fine tuning of a pig production system can be much more complex than that of a motor engine, since not only are the principles of physics and chemistry involved, but so also are the very complex laws of biology, involving as they do the delicate interactions of the pig with a host of influential factors, such as nutrition, climate and micro-organisms. The stockperson has a crucial role to play in controlling these influential factors for the benefit of the animals in his care and, of course, one of the most important of these influential factors is the relationship he himself is able to develop with his stock, as discussed in Chapter 4.

Perception of the need for skills and training

In a recent study on pig farms in Canada, employers and employees were asked to state if skills were required to perform particular jobs on the farm. A higher proportion of employees than employers acknowledged the need for skills to perform each of five jobs listed (see Table 8.1). This response of employers can be taken as indicating their tendency to take their own skills for granted. The employers and employees on these pig farms were also asked for their opinion as to which jobs in the list required training. Somewhat conversely, more employers (approximately 50%) than employees (approximately 40%) considered that training was required for the five skills listed. This latter finding may indicate that many employees also have a tendency to undervalue the skills which they have acquired to do the various jobs on the farm. Also worthy of note is the relatively low proportion of both employers and employees who acknowledged a need for training in skills. On farms in which such training

Table 8.1 **Employer/employee evaluation of skills required to perform various jobs and of the skills which required training**

	Employers	Employees
Jobs	*% who considered that skills were required to perform the job*	
Pig feeding	87	95
Sow breeding	66	75
Farrowing	66	77
Weaner management	66	78
Building and barn maintenance	71	84
	% who considered that training was required to develop the necessary skills	
Pig feeding	60	43
Sow breeding	48	37
Farrowing	46	37
Weaner management	49	37
Building and barn maintenance	45	36
Estimate of total man days training provided in a year	241	114

Source: Based on Howard *et al.* (1990)

was provided, employers considered that many more days (241) had been devoted to training in skills than stockpeople thought they had received (114 days). Thus there appeared to be much confusion as to what constituted training.

Training needs

In the Canadian study cited above, a very high proportion of both employees (98%) and employers (95%) were satisfied with the training provided/received. Lloyd (1975) was much less satisfied with the level of training of personnel in his study of the poultry industry. Lloyd cited an Agricultural Training Board study carried out in 1974 of manpower and training needs in the British Poultry Industry which established that only 1% of the workforce had poultry qualifications. In addition, although 86% of the workforce had no experience of poultry before their present job, only 4% had attended short training courses in the 2 years prior to the study. Lloyd concluded that the poultry industry was either attempting to train its workers by example or else relying on an untrained workforce with the removal of the skill element from the job through system modifications and automation. He concluded that such management policies were unlikely to be effective and cited in support of this contention the general state of job dissatisfaction of many workers. This was reflected in high labour turnover (1 out of 4 workers leaving each year), recruitment problems as well as inferior work.

Thus Lloyd contended that training of poultry workers to improve their understanding of the animals and to upgrade their technical skills would not only enhance animal care but would also have positive influences on job satisfaction, work performance, employment stability and the ease of recruitment of new staff.

More recent evidence indicates that the pig industry in Britain does not pay sufficient attention to training. In a study of 65 pig stockpeople from 15 farms in Scotland, it emerged that 86.8% of farms studied had no 'off-farm' training scheme for stockpeople and that 53.3% had no 'on-farm' training scheme (Segundo, 1989). In addition, 34% of the stockpeople expressed dissatisfaction with the training opportunities offered by these farms.

It would appear that pig stockpeople in Australia are no better trained than those in Britain. Cleary (1990) cites studies carried out in the 1980s on pig farms where only about 1 in 8 stockpeople had some element of training including brief short courses and technical seminars. Thus a general inadequacy in the state of staff training was identified. This was considered to be a serious state of affairs in view of tightening financial margins and increasing technological complexity in the industry. The lack of adequate training opportunities for pig farm staff was not unconnected with the high staff turnover in the industry.

Requirements of stockpeople for training

Stockpeople require training on 3 fronts:

- skills associated with the job
- increased understanding of the animal
- improved attitude towards the animals, the production system and the business goals

SKILLS

Stockmanship skills include general handling techniques for catching, restraining and moving pigs as well as the more specific skills required in such tasks as injecting, rectal temperature recording, pregnancy diagnosis, assisting farrowing, weighing, ultrasonic measurement of backfat thickness, assisting at mating, collecting/processing/examining/evaluating semen, carrying out artificial insemination, applying identification marks/tags to pigs, operating cleaning equipment and adjusting environmental control devices.

Other stockperson attributes include perceptual skills. These incorporate the senses of sight, hearing and smell which are so vital to the stockperson's observations in assessing normality and in detecting the early signs of departures from the normal so that problems are detected and rectified in the early stages of development. Recording and data collection skills are also most important so that management can monitor actual performance (both physical and financial) against budgeted targets and against the achievements of other (e.g. top 10%, top third and average) producers.

UNDERSTANDING THE PIG

The better the understanding stockpeople have of the pig and how it operates, the more sensitive they can be in providing for all its needs.

Thus, increased awareness of the structure, physiology and behaviour of the pig and of its requirements in terms of nutrition, climatic environment, housing and maintenance of good health is of crucial importance to the enhancement of stockmanship. Increased understanding of the animal stimulates greater interest in it on the part of stockpeople and motivates them to renewed efforts in tending the stock in their care.

ATTITUDE TOWARDS THE ANIMALS, PRODUCTION SYSTEM AND BUSINESS GOALS

Another role for training is to strengthen the bond between a stockperson and the animals being cared for and to enhance the connection between the business organisation (especially its management) and the stockperson (as its employee).

A part of each training programme can have the explicit goal of influencing personal attitude to the animals so as to improve their care. This component of the training session can be prepared on the basis of certain principles:

- We all have it in us to be cruel and violent both to our fellow man and to animals in our care, albeit on rare occasions for the vast majority of people. This potential problem in all of us can be dealt with by its open recognition during training.
- Although pigs are unable to report of 'hurt feelings', inflicted pain and cruelty, their behaviour can be an indicator of kindness (or unkindness). If the dog clearly displays wounded pride after a scolding from its handler, the pig is capable of similar responses although we are not so competent in recognising this reaction in the pig.
- Certain things trigger unkindness and cruelty in stockmanship. Aggression is more likely when frustration is caused by both domestic and work related factors and a solution seems improbable in the short term. The training programme must contain a section on lines of communication between new recruits, less experienced staff, section managers, unit manager and owner so that problems causing frustration and distress can be aired and hopefully resolved quickly before maltreatment of the animals becomes one of the safety-valve outlets for such pent-up emotions.
- Kindness towards animals in care can be instilled and developed in the carers/stockpeople. The training programme must contain

a clear indication that management approves and expects 'talking to' and 'touching/stroking' the pigs as an inherent component of good husbandry.

- Management has responsibility for the correct attitude of all involved towards the animals. The training programme must contain mention of 'kindness to and care of pigs' as a business goal.

Training methods

The training of stockpeople can take a variety of forms:

- an apprenticeship working along with an experienced stock-person
- informal in-house instruction from and/or discussions with the manager/owner and other staff
- formal in-house training sessions
- industry seminars
- external training on day or block release courses
- external training in the form of full-time specialist courses on pig production and pig handling skills lasting from a few months to one year
- reading of literature on the pig and on pig production

The traditional training for the pig industry was the son 'apprenticed to' his father or a new entrant employee working with and under the guidance of an experienced stockperson who may also have been the farmer. The effectiveness of such training was dependent not only on the competence of the trainer as a stockman but also on his ability as a teacher to pass on his knowledge and skills to the trainee. Of course, the effectiveness of the teaching was also dependent on the interest, enthusiasm and intelligence of the apprentice.

According to Cleary (1989), discussions between experienced staff and newcomers still form the most frequent training activity on most pig farms in Australia. He contends that, while such discussions can be very useful, an external information source should also be used regularly to stimulate discussion. Such sources could be a pig industry journal, a private consultant or a government advisory officer. Other forms of informal training found useful by Cleary were 'trade nights' which provide information on commercial products, discussion group meetings and seminars on relevant research findings. Owners of pig farms who encouraged their staff to attend such functions were rewarded with lower staff turnover rates, higher productivity and improved job satisfaction.

ON-FARM TRAINING

A study in Scotland (Wright, 1985) indicated that pig stockpeople preferred on-farm training to other types of training. On-farm training schemes have the advantage that they can be purpose-built to meet the needs of each farm. Often managers and staff are too familiar with their own farm as it stands and fail to look at the situation objectively. It can take a pair of expert eyes from outside to see all components of the farm more closely, to spot the weaknesses and to focus on these in the training provided. If an outside instructor is used on such exercises it is most important that he/she should form a partnership with the farmer/manager in presenting the training course. This initial and continuing partnership will help the manager to provide continuing training for his staff.

Stockpeople undertaking a two hour on-farm training course in the early afternoon of a normal working day.

First the instructor should convince the manager about problems that exist and the instructor thereafter can depend on the manager to consolidate his messages to the staff. The staff, on their part, should be encouraged to participate fully in discussion and their experiences, observations and suggestions should be carefully noted and incorporated into the amended policy for the farm. In this way, manager and staff will act as a team and be fully involved together in decision making and in the formulation of amended procedures. This has an important influence on each staff member identifying with the objectives of the enterprise and thereafter with their resolve to work towards the attainment of these objectives.

This in turn gives such in-house training an advantage over external training courses, for all relevant staff can attend the course and any confusion over recommendations can be resolved by discussion among all those present. On the other hand, a stockperson attending an external course will not find that all the material presented is relevant to his unit. He may also bring home an idea, be enthusiastic about it himself but receive only lukewarm interest from his employer and colleagues because they foresee difficulties, for one reason or another, in applying the idea on their farm. The employee who attended the course may either have wrongly picked up the message from the course or else may lack the ability and the confidence to convince colleagues that the idea is worthy of implementation on their farm.

On-farm training can also be made more available to all employees by fitting it into an appropriate part of a working day, paying overtime as necessary if the working day ends up longer than usual.

Another advantage of in-house training is that it eliminates health risks associated with personnel from different farms meeting, for example, at a day release course at a central location.

EXTERNAL TRAINING COURSES

Formal external training courses suffer from the disadvantages outlined above relative to in-house courses. However, they can be extremely useful for the training of stockpeople in particular skills e.g. collection, examination and processing of semen followed by artificial insemination. They can also be invaluable in the training of managers in such aspects as human resource management, as well as in an advanced understanding of the anatomy, physiology and behaviour of the pig and of its overall requirements for health, welfare and production efficiency.

Inexperienced young people aiming to become competent pig

stockpeople can obtain effective introductory training in the skills and knowledge required for the job. Of course, such initial training must be supplemented regularly with effective 'in-service' training when they obtain a job as a stockperson. Beynon (1990), in his capacity as Head of the Department of Agriculture and Farms Director at the Berkshire College of Agriculture in England, comments on the availability of specialist full-time block release, day release and short courses to provide training in either pig stockmanship or the management of stockmanship. He points out that the success of such college courses is linked to the commercial pig performance measures achieved on the college farm and the relevance of the pig systems in operation to those operating commercially. He points out further that the personality of the college pig stockperson is crucial in attracting, motivating and developing stockmanship skills in full-time and part-time students.

Evaluation of the effect of training on understanding and skills

A training programme will incur costs for the business, so the owner/manager will want the training to provide benefits which will at least cover these costs. A training programme can range from being very effective, fulfilling all its objectives, to being totally ineffective. Thus the impact of training on 1) the improved understanding of the pig and of how its needs can be met more effectively, 2) skills, 3) interest level in the job and 4) motivation for the work must be evaluated. This will help to formulate a basis for future training in terms of form, location and frequency. As Cleary points out, reviewing the effects of training with the participants helps to make the process 'constructive, effective and maintains the trainee's enthusiasm.'

The evaluation of the impact of training can be based on a combination of open discussion with the managers and staff (in which the staff are stimulated to participate fully) and by a questionnaire which all participants should complete. A short 'fill in the missing word or number' and/or 'true or false' series of questions can help to establish how much of the material presented and the ensuing discussion was understood and put into the context of operating their own pig production system by the participants.

Of course, in order to determine objectively how much has been gleaned from a training session it is imperative that the level of knowledge and appreciation of the material to be covered during a training session is evaluated before as well as following the course. In addition, this pre-training test can be used to indicate to the trainee the subject matter to be covered in the training programme and the key points to be noted.

Another important role for on-farm training is in the caring and helpful induction of new employees. Whether or not such employees have had previous training and/or experience, they require careful initiation into the system and procedures operating on the new farm. Providing such training and awareness from the outset is most important in relation to making them feel welcome and involved while increasing their confidence, interest, job satisfaction and loyalty to the organisation.

According to Cleary, in Australia poor induction procedures for young recruits contribute to the fact that labour turnover rates are highest among workers with only a few months' service. Thus, throwing new entrants carelessly in at the deep end 'to learn for themselves' is totally counter-productive and results in the industry losing potentially good recruits.

Stockpeople undertaking a simple test of their knowledge and awareness prior to an on-farm training course. Such tests are held before and after a training course as one means of determining the effectiveness of the course.

While the impact of training on an improved understanding of the pig and its needs can be tested in the manner outlined above, the influence of training on skill level and competence in carrying out the range of jobs required is more difficult to assess. These characteristics are probably best assessed by the unit manager or senior stockpeople who are working with the younger trainees. Fairly successful techniques to measure the effect of training to improve understanding and handling skills on the level of empathy the stockperson has with the pig have been developed by Hemsworth and his colleagues in Australia. The approach behaviour of the pig towards a stockperson in controlled conditions is used to evaluate this most important characteristic in stockpersons. Details have been outlined in Chapter 3.

On-farm training in practice

A pilot study of on-farm training was carried out by the University of Aberdeen (Beveridge and colleagues 1992). The study involved 3 large pig farms and a total of 20 workers. The stockpeople involved varied widely in age, length of time working with pigs and in the amount of training in pig biology and pig handling skills received previously at college courses, at day release classes and at short off-farm and on-farm specialist training sessions.

Training courses, each lasting for 2 hours, were conducted in the farm office on the following subject areas:

A. Management of the sow and piglet during farrowing, lactation and immediately post-weaning.
B. Pre-mating, mating and post-mating management including AI.

Each training session involved:

1. Presentation of slides featuring aspects of pig behaviour and appearance along with relevant components and weaknesses of the system operating on each farm in turn along with other slides illustrating important pig husbandry principles.
2. Discussion.
3. Assessment of each stockperson's understanding of pig biology and the basis of various skills related to the care and wellbeing of the pig. This was done by asking each participant to provide written answers to short answer questions immediately before and immediately after each training session.

The results of these tests are presented in Table 8.2.

Table 8.2 Results of pre- and post-training tests (written responses to short answer questions)

	Farrowing unit staff			Service/pregnancy area staff		
Farm	A	B	C	D	E	F
Pre-training score (%)	*	61	64	*	63	60
Post-training score (%)	75	68	74	85	73	80

* No pre-test given to this training group

On the two farms on which 'before and after' tests were carried out there was an increase of 19% in the average post-test score (74%), relative to the pre-test score (62%). As might be expected, higher average test scores were attained by older stockpeople, those with longer experience with pigs and by those with more previous training on the pig. However, a few stockpeople with little work experience with pigs showed large increases from pre- to post-test scores.

The stockpersons' general assessment of the training courses provided and their opinions on training in general were sought approximately one week after completion of the training sessions by personnel other than the trainer. The results of the overall assessment of the training courses are presented in Table 8.3.

Thus, stockpeople were generally satisfied with the aspects of course content and presentation. In an attempt to gauge the attitude to training of the stockpeople involved they were asked to express their degree of support (or lack of support) for 6 prepared statements which related to on-the-job skills, understanding the pig and the relevance of training.

The statements receiving least and most support were the following:

Table 8.3 Evaluation of aspects of the training courses provided

Aspect	Average score (Scale: 1 = poor, 5 = very good)
Clarity of speech	4.7
Relevance of audiovisual aids	4.6
Ease of understanding of course contents	4.2
Pace of delivery	4.1

Least support: 'Stockpeople currently employed in the pig industry have an adequate level of skill'

Most support: 'Training will improve my potential as a pig stock-person'

Thus, a very clear need for training was expressed.

The responses to other questions established the following:

1. A unanimous desire to have 'tests' carried out before and after the training sessions.
2. A decided preference for regular (e.g. once a month) training as against training at irregular and infrequent intervals.
3. A strong preference for on-farm relative to off-farm courses.
4. A strong preference for the working team to be trained together.

Probably the most surprising outcome of this exercise was the desire of stockpeople to have their understanding and the basis of their skills evaluated by testing. Obviously there was a strong desire to become better at their job and they saw regular on-the-job training and evaluation as the best means to that end.

While this was a small pilot study, it helps to provide some guidelines on:

- determining training needs in other situations
- the most effective methods for providing training according to the requirements of stockpeople (either as individuals or as teams)
- how to evaluate the relevance and usefulness of such training
- evaluating stockpeople on the basis of their understanding of the pig and on the basis of many of their skills
- the progressive development of training
- components of the criteria for the award of progressive attainment qualifications to stockpeople

Some of these important aspects are developed further later in this chapter.

Training in pig-handling skills

Effective training programmes have been developed to train personnel for many specific stock tasks such as injecting, castration, mating management and supervision, artificial insemination, pregnancy diagnosis, body condition scoring of sows and ultrasonic measurement of backfat on the live animal.

However, the handling skills and piggery conditions required for simple operations such as moving pigs from A to B have not, to date, been given their due amount of attention in training programmes. There is a considerable amount of such stock movement in the day-to-day operation of a pig enterprise as sows are moved to the farrowing house, weaned sows to the breeding/mating area, weaned piglets to the nurseries, growing-finishing pigs to the successive stages in this operation and, finally, finished pigs to the collection/loading area before directing on to the transportation vehicle. If problems are encountered in moving pigs there are several undesirable consequences. Much extra time and physical effort on the part of the stockpeople may be required to complete the operation. In addition, great frustration and bad temper can be generated in the stockpeople as the pigs stall, deviate from the intended path and even turn around; distress and even physical injury to the pigs can arise as a consequence. Such damaging influences on stockpeople and pigs can be avoided by a little prior planning and Hemsworth and his colleagues in Australia have given much thought to the ease and safety of movement of pigs.

Some of the simple guidelines they specify are the following:

- Plan the buildings effectively
  (a) Avoid making the main access passages too narrow.
  (b) Ensure passages are bounded by solid walls since then pigs being moved will not be distracted by their fellows in pens adjacent to the pass.
  (c) Provide a see-through gate at the end of the corridor since pigs will move more readily towards such than towards a solid gate or door.
  (d) Ensure passages are uniformly well lit since variations in lighting from starting to end point can disorientate pigs.
  (e) Ensure even floor surfaces since pigs can be distracted and distressed by uneven surfaces.
- Clear the way in advance
  (a) Ensure that there are no obstacles in the passage way. Pigs may be afraid of such strange objects or just curious and some will stop to nose, smell and investigate and so disrupt a good flowing movement of the batch.
  (b) Avoid any food spillage or water in the passageway as this will be distracting. Some pigs will stop and hold up their followers, a pig may turn round to be followed by others thus wasting time and causing frustration to the staff.

- Careful handling
  - (a) Do not attempt to move an excessive number at a time in relation to the width of the passage.
  - (b) Always use a solid stockboard behind the pigs to present a solid wall between the stockperson and the pigs.
  - (c) Do not use electric goads or sticks to move pigs. Pigs should not be slapped (even 'gentle' slapping over a period will increase a pig's level of chronic stress and fear of humans). Handling should be based on gently rubbing, stroking and 'talking to' the pig—so-called 'positive handling techniques'. This reduces the pig's fear of humans, increases its confidence in the stockpeople and makes it easier to handle.
  - (d) Avoid loading ramps which are too steep. Provision of lifts to load and off-load pigs are very desirable in relation to minimising frustration to staff, distress and injury to pigs and subsequent meat quality problems.

Impact of training & interactive staff-management discussions on motivation and job satisfaction

Not only must training develop understanding and skills but it must raise the level of interest in the job and job satisfaction. Such benefits will arise from training if it is well organised, carried out by competent and enthusiastic leaders, if trainees are stimulated to participate fully and present their own ideas and experiences and if their needs and ideas are used to formulate the basis of subsequent courses.

The attitude of the farmer/manager towards training and during its execution is most crucial. If he/she dominates the session, instructing staff what to do and how to do it, without stimulating discussion among staff and using such feedback effectively in formulating amended policies and working routines on the farm, as well as in the planning of future training, then the time and effort devoted to such so-called 'training' is likely to be largely wasted.

On the other hand, when training is well thought out and executed it is the single most useful tool to improve the performance of stockpeople and probably one of the cheapest approaches to improving the farm performance in general. This principle might be partly masked by the ambiguous concept most people have of 'training'.

Classical training has mainly been, until now, a one-way, passive education process, or in some cases (such as instruction in the use of pregnancy detectors and AI) it has constituted short 'training in skills' courses. Such one-way passive training strategies, although often adequate for specific purposes, have neglected some of the major principles for the improvement of staff performance—notably that of staff motivation.

An extension of the concept of training towards participative 'two-way' training and discussion sessions involves the following key aspects:

1. Evaluation of further training needs. Training in skills (such as oestrus checking, AI, farrowing assistance and fostering) may or may not be a real need on a particular farm, but training can also be directed to improving the general understanding of 'how things work', which in turn stimulates interest. An example of this was given by a well known specialist pig veterinary surgeon, who dedicated time to informal chats with stockpeople, during which he discovered that those on a particular farm had the desire to understand more about the pig's anatomy and physiology, and to comprehend the mechanisms involved in diseases such as piglet scouring (Muirhead, 1983). After a series of on-farm presentations and discussions, farm production improved significantly. This was thought to be because of a better understanding (and better decision taking) by the stockmen and because involvement and motivation were fostered.

   The lesson from this example might be: decide what training you consider staff require, but also consult staff to see what they want to learn about. For this, a form with various options for training can be completed by staff and later forwarded to those who are to be involved in training.
2. Staff involvement. This stimulates the sense of participation and belonging and also the concern for achievement of objectives. Why should staff be interested in being trained, if they are not made to feel an active part of the process of achieving the farm's targets?
3. Goal setting and monitoring. When staff are involved in the target setting of the company and in the analysis and monitoring of results (against targets), and management/production methods are clearly understood by all the staff involved, interest and motivation are bound to rise. The informal discussion generated during this activity might well be a most valuable source of informal training. This can also be seen as the best opportunity to detect 'training needs'.

4. Feedback and reward. Feedback from staff is not only the best way to learn about their feelings and skills in relation to their work (which in turn reduces staff related problems), but it can also be a useful source of knowledge for improving the system, since a formidable number of working and observational hours and experiences have been accumulated by the collective staff. Most of all, making them feel part of the system might be the single strongest source of potential motivation.

   Reward can have two basic components: the recognition of achievements is very important (Segundo, 1989) but also, particularly in relatively low paid jobs, incentive schemes such as special prizes or awards or bonuses should be considered.
5. Overall evaluation of training. While improved performance should result from training, one should be fully aware of the wide range of influences that could contribute to the overall result. Success might be quantified in terms of better productivity, higher quality of output, less absenteeism, lower staff turnover, greater adaptability, fewer accidents, less wastage of resources and lower costs of maintenance of equipment. According to Bennett (1989) 'Training can improve workers' morale, create better interpersonal relationships, instil in employees a sense of loyalty to the organisation and provide other intangible benefits.' However, the training courses must be evaluated by managers and employees alike in order to improve and expand them and to determine aspects such as optimum location, duration and frequency.

Stimulating participation of staff

Time spent in training sessions should be considered as overtime by management and paid as such. Management should provide simple refreshments (e.g. coffee and biscuits) for every session. Regular staff training and re-training meetings must be presented and prepared as important events.

Attendance should be recorded and the opinions of both managers and stockpeople of the usefulness of these meetings should be requested. During a study in Scotland it was found that some managers had been unsuccessful in achieving good attendance and/or participation of staff in such training sessions. It is likely that in most cases, although not all, interest will increase if the stockperson can perceive that he or she will obtain some benefit from such training activities. Paying overtime for the time spent on training might be the first incentive (particularly for low paid workers) which may result in the discussion/training session enhancing the interest of such low paid and/or poorly motivated staff.

Determining retraining needs

Motivation, like theoretical knowledge, fades away with time so that training and discussion sessions should not be a 'once in a blue moon event'. The need for and frequency of a refresher course and staff/management meetings must be assessed by various methods and will be determined by various factors. Changes in staff, drops in productivity, frictions within staff, or even generally poor morale due to low pig prices or other adverse circumstances, may influence the need for bringing forward planned discussion and training sessions. A full evaluation of the previous course (see suggested evaluation sheet in Table 8.4) should provide information as to which areas to tackle next, how training sessions can be made more attractive, and what the optimum frequency of meetings might be. The most important features of these training sessions which must remain alive in all future meetings are: Discussion, Analysis, Motivation, Involvement and Training (i.e. the D.A.M.I.T. strategy). Variation in form according to needs, discussion of future plans and previous experiences are basic to keeping the interest in these meetings alive. As the old saying so rightly goes, 'variety is the spice of life'.

In-house meetings and training

Outside advisers invited to some of the within-company staff/ management meetings can stimulate discussion, act as a moderator, contribute using personal experiences (informal training), help to detect training needs and generally maintain the interest in these sessions.

A frequency of stockpeople/management meetings at approximately 3 monthly intervals is probably appropriate in most situations but this might not necessarily reflect the demands for training. So, quite apart from the stockpeople/management 'trouble shooting' or 'stock-taking' discussion meetings, with or without an invited adviser, specially designed training sessions must be organised at a frequency related to the assessed training needs of the farm.

An outline of the 'D.A.M.I.T.' workshop training course

The authors have evolved a training course named 'D.A.M.I.T.' (Discussion, Analysis, Motivation, Involvement and Training) which, as its name suggests, implies a series of different meetings designed to: 1) Analyse and improve (where possible) stockpeople management and performance, 2) analyse and improve (where possible) the management structure, 3) analyse and improve (where possible) the farm's performance.

The basic approach is as follows:

Step 1. A farm visit and analysis by the adviser.
Step 2. Formal training sessions with managers, critically examining particular aspects of personnel management.
Step 3. Stockpeople-managers-advisers' discussions about farm performance and problems and to formulate policies including arrangements for regular and progressive training in 'understanding and skills' sessions.
Step 4. Regular 'understanding and skills' training sessions.

The sessions are specially targeted to change attitudes of stockpeople and managers, bringing them closer together in decision making and in the daily operation of the farm. The adviser not only stimulates input from staff, acting as a facilitator and moderator in this role, but also contributes with his own expertise in these open sessions.

EVALUATION OF THE WORKSHOP

In the last session of the DAMIT course the whole workshop is evaluated. The main aim is to obtain useful feedback from stockpeople and managers in order to optimise the development and improvement of further courses. An example of the evaluation sheet used is presented in Table 8.4.

It is always advisable to evaluate how much was learned in the course. A short 'fill in the blanks' and/or 'true or false' type questionnaire can be developed to test the effectiveness by which the theoretical and practical knowledge was transferred to and absorbed by the staff.

In the study cited by McCormick & Ilgen (1982), training groups which received feedback about their results after a training session

111

Table 8.4   Evaluation sheet for a DAMIT workshop training course

---

EVALUATION

Pig Stockmanship Training Course

We want to find out what you thought about this training course and also to get your ideas on further training.

Your constructive criticisms of the training course you have just completed together with your views on stockmanship training in general will help us to improve on training in future courses. We would be grateful if you could complete this short questionnaire for us before leaving. There is no need for you to put your name on it.

PART ONE   The training course just completed
1. The following statements refer to this training course. Please indicate your opinion regarding the extent to which you agree or disagree with each statement.

Statement	Strongly agree	Agree	Neither agree nor disagree	Disagree	Strongly disagree
a) My knowledge of the needs of the pig and of pig husbandry has improved	☐	☐	☐	☐	☐
b) The content of the course was relevant to my work with pigs	☐	☐	☐	☐	☐
c) Overall, the course was stimulating	☐	☐	☐	☐	☐
d) I felt that the course was pitched at the right level	☐	☐	☐	☐	☐
e) My animal caring skills will be better as a result of this training course	☐	☐	☐	☐	☐

2. We would like to know how you rated this course.
    The scale is 1 = very poor, 2 = poor, 3 = OK, 4 = good, 5 = very good

    (Please circle the appropriate number)

	Very poor	Poor	OK	Good	Very good
A. *Presentation*					
a) Use of visual aids	1	2	3	4	5
b) Encouragement to make comments and suggestions	1	2	3	4	5
c) Opportunity to ask questions	1	2	3	4	5
d) Handouts for reading after the course	1	2	3	4	5
e) Overall organisation	1	2	3	4	5
B. *The Presenter*					
a) Clarity of speech	1	2	3	4	5
b) Legibility of handwriting	1	2	3	4	5
c) Helpfulness when dealing with queries	1	2	3	4	5
d) Pace at which material was covered	1	2	3	4	5
e) Overall style of teaching	1	2	3	4	5

3. Did you find this course   Please tick
    Very difficult   ☐
    Difficult   ☐
    Easy   ☐
    Very easy   ☐

PART TWO   Training in general

How often would you like training to take place
    Once a month   ☐
    Once every three months   ☐
    Once every six months   ☐
    Once a year   ☐
    Not at all   ☐

How important do you think it is to keep on having courses like this
    Not at all   ☐
    Slightly important   ☐
    Fairly important   ☐
    Very important   ☐
    Extremely important   ☐

Please suggest a topic (or topics) you would like to have covered in any future training course. _____

achieved higher average performance gains over a period of training than groups which received no feedback.

Recognition of attainments following training

The improved understanding of the pig and enhancement of skills emanating from effective training has positive influences on the dedicated stockperson's interest in the job, on motivation and job satisfaction. In turn the wellbeing of the pigs, their performance and the profitability of the enterprise should improve. While stockpeople get further satisfaction from seeing such improvements, it is important that their higher level of attainment following training is recognised in some formal way.

Currently in Britain the National Council for Vocational Qualification (NCVQ) has initiated a system of Competence Assessment based in the workplace across all industries. 'Competence' simply means 'the ability to do the job or task at the defined level with a high level of repeatability.' The scheme requires that all employers and supervisors be formally trained as so-called 'primary assessors'. Their assessment of 'competence level' will in turn be verified by trained verifiers. This scheme could provide a useful structure for progressive training of pig stockpeople and development of stockmanship skills through various levels of competence. Each person participating in the Scheme will have a N.R.O.V.A. (National Record of Vocational Achievement) and will have an N.V.Q. (National Vocational Qualification) at any level from 1 to 5 as defined below:

Level 1—a range of work activities which are primarily routine and predictable or provide a broad foundation.

Level 2—a broader and more demanding range of work activities involving greater individual responsibility.

Level 3—skilled areas that involve performance of a broad range of work activities including many that are complex and non-routine. Supervisory competence may be a requirement at this level.

Level 4—the performance of complex, technical and professional work activities, including supervision and management.

Level 5—at the time of writing (early 1992) this is under development and is the subject of consultation with professional bodies and other interests.

Relationships between man and animals—some past and present examples from everyday life

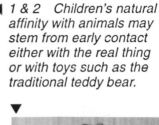

◀ *1 & 2 Children's natural affinity with animals may stem from early contact either with the real thing or with toys such as the traditional teddy bear.*

▼

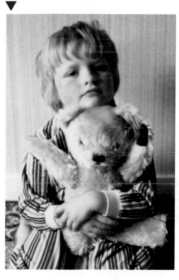

3 Most people's best appreciation of man–animal empathy is derived from the natural relationship with family pets. Shambles the Beagle— boisterous, free-spirited, exasperating and wholly lovable. His coming brightened and enlivened our days and his going broke our hearts.

▼

4 General early morning activity and togetherness in the poultry house.

▼

▲

5   In many areas where cows grazed on open common land during the day, it was usual practice for the milker to go out just before milking time was due to call the cows home to the byre or milking shed. The traditional call in the West Coast of Scotland was 'Trodhad' (Gaelic for 'Come here'). On hearing the call, the cows, which were usually out of sight of the milker, would cease grazing and begin the homeward trek. In winter, their homecoming might be influenced by the anticipation of receiving supplementary feed, but they responded to the call equally well in summer when no feed was offered on entry to the byre for milking. This is a good example of leading rather than driving stock.

◀ 6   The stockman checks one of his steers which had been off colour, showing little interest in feed on the previous day.

7   Are you yet another of these Aberdeen University students? At least you are a better communicator than some of the rest of them! (Photograph by Marysia Stamm.)

▼

▲

8   The operation of transporting sheep in open boats for summering to offshore islands on the West Coast of Scotland reflects many aspects of skilled stockmanship including empathy, shepherding and handling skills during loading, transporting and unloading. (*Changing Pastures*, painted by Rosa Bonheur. City of Aberdeen Art Gallery and Museums Collections.)

9   Similar stockmanship skills. Dutch stockmen using a punt to transport their dairy cows one at a time to new pastures across the canal. (Netherlands Ministry of Agriculture, Nature Management, and Fisheries. Information and External Relations Department.)

▼

▲
10   The shepherd and his dog at work with hill sheep above Balmoral
Castle, Deeside, Scotland. (Painted by John Mitchell. City of Aberdeen Art
Gallery and Museums Collections.)

11   The faithful collie concentrates on the sheep as the shepherd (right)
chats about hill farming, the weather and life in general.
▼

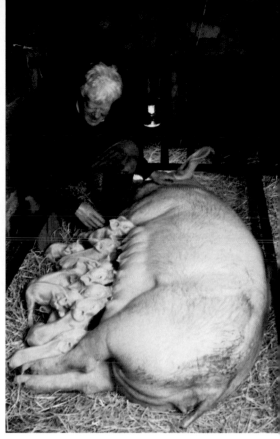

12 & 13   Massaging the udder of the sow (above) to check for the presence of pre-farrowing milk. Not only does the sow appear to enjoy such attention as she rotates her trunk to expose more of her udder, but this also helps the good stockperson to develop an even closer rapport with the sow. By the time she farrows she has increased trust in the attentions of the stockman when he checks progress, assists farrowing as necessary and (right) helps the weaker piglets to get to the udder for their due share of colostrum.

14   The stockman and his pigs getting to know each other better on an outdoor system.

Caring and sensitive stockmanship, in achieving a better man–animal relationship, has a major effect on welfare, health, reproduction and growth in pigs

15 To achieve maximum success with AI, close contact with the boar in the adjacent pen is desirable as well as careful and pleasant handling by the stockman.

16 Cecil Morris carries out pregnancy diagnosis in a group of very peaceful sows.

17 As Ken Rae checks the wellbeing of his weaners, they in turn communicate with him.

In terms of improving the understanding and technical skills of the stockperson, on-farm training, involving interaction with and handling of the animals (as well as sessions in the farm office), has much to commend it

18 Stockpeople receiving training in measurement of backfat in gilts and in selection of replacement gilts. ▶

19 Young stockpeople being trained to body condition score sows. Backfat is measured ultrasonically so that individuals can relate their condition scores to the actual backfat thickness. As their scores, with practice, become increasingly more closely correlated with actual fatness, so their confidence in body condition scoring increases. ▶

20 Interacting with and discussing very tame Sus scrofa. ▶

Improving stockpeople's understanding of their animals

◀ 21 A group of managers and stockpeople on a large pig farm in Colombia prior to an in-house training course.

◀ 22 A training course for stockpeople in Mexico, conducted by Juan José Maqueda on the right.

◀ 23 Stockperson Louise Addison receiving a prize for achieving very high levels of piglet care and survival at the annual awards presentation of a large pig breeding company in Britain. At the same time Louise received an Agricultural Training Board Award. (Pig Improvement Company Journal.)

Providing better opportunities and working conditions for stockpeople will improve job satisfaction and the motivation for better animal care

24   Trade nights which provide information on commercial products and discussion group meetings and seminars which inform people about research are other means of providing informal training. Cleary in Australia found that owners/managers of pig farms who encouraged their staff to attend such functions are rewarded with lower staff turnover rates, higher productivity and improved job satisfaction. The picture shows Dr Paul Hemsworth having a lively discussion with pig farmers, managers and stockpeople in the north-east of Scotland.

25   The cooking/dining/resting/recreation room in a large pig enterprise in Chile (Super Pollo Company, managed by Dr Gonzalo Castro). While their lunch is cooking, two of the staff relax over a game of cards.

26   In addition to welcoming new recruits, it is equally important to look after the welfare of retired employees who have given skilled and loyal service. A retired employee of the Grupo Delta Company in Mexico relaxes at the entrance of the farm where, during the working day, he keeps a check on movement of pigs, feed and personnel, thus boosting his retirement pension and retaining contact with his former colleagues.

Stockmanship and man–animal empathy can involve a three-way relationship

▲

◀ *27 & 28   Special skills, as well as great empathy and patience, are required in training sheepdogs so that the dog controls the sheep by persuasion to move in the required direction rather than by force. The photograph above shows shepherds and dogs persuading their flock to cross the fast-flowing Jollie River on Braemar Station in the Mackenzie Country of South Island, New Zealand. The shepherd and dogs must, with great patience, win the battle of wills to get the leading sheep to start crossing the stream, as then the remainder will follow.* (Photograph: Andris Apse, New Zealand.) *The photograph to the left shows the shepherd cradling lambs which appear to be very reassured and secure in his arms, while the well disciplined dogs at heel concentrate, among other things, on other sectors of the flock.* (Reproduced with the kind permission of Barclays Bank Agricultural Services Department.)

29   The particular relationship of horse and rider reflects mutual emotional attachment and confidence of the partners, sensitive and effective communication, and pleasure in each other's company and in joint achievement.

30   There was 'method' in the traditional dairy maid singing to the cow during the milking process in fairy tales and in real life. There was probably a great deal of vocal communication and 'praise with the voice' to encourage the foraging cow to stop for milking. Once the milking started, the rhythm of the singing would coincide with that of the milking process. This was effective stockperson–animal communication—perhaps more especially if the cow could appreciate a fine singing voice—and contributed to the cow standing at peace during milking, good milk 'let-down' and efficient milking. (Photograph reproduced by kind permission of National Museums of Scotland.)

31   The Syrian shepherd leads his sheep to new pastures while his fellow nomads flank the flock at the ready to protect the sheep from oncoming traffic. (Photograph: Janet Roden.)

▲

32   A Mexican smallholder leading his cattle and goats to graze the available herbage alongside a busy modern motorway.

◄ *33   The 9 month old red deer stag Broinean (Gaelic for 'puny little fellow') and his carer Alan communicate with each other. Broinean was found as a tiny starving orphan, bottle-reared by Alan's wife Marnie and spent his first winter sharing the shelter and feed with the sheep and cattle. Perhaps now, as spring is phasing into summer, Broinean is considering a reunion with his own kind in the wild.*

35   Nadia 'supervises' the preparation of part of this book.

▼

34   Anne and Polly catch up with each other's news of the day.

▼

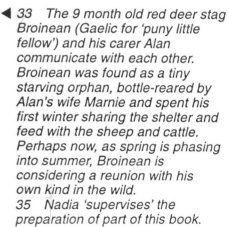

These competence or 'can do' levels are not designed to represent academic qualifications but nevertheless they have been equated with different levels of academic attainment at school, college and university. For example Level 4 is approximately equivalent to Higher National Certificate or Pass Degree Standard while Level 5 is approximately equivalent to an Honours Degree.

The British Government has set the following targets (among others) for N.V.Q.s across all industries:

- *By the end of 1992* no young person should be employed in a job without training; two thirds of the workforce should have achieved an N.V.Q. at Level 2 or its academic equivalent while at least a quarter should have reached Level 3.
- *By the year 2000* a minimum of at least half of the employed workforce should be qualified to Level 3 or its academic equivalent (an advanced vocational qualification comparable to an A level).

Beynon (1991), while appreciating the great merit of recognising different levels of 'competence' among stockpeople, has expressed some uncertainty about the likely impact the above scheme will have in practice for stockpeople in livestock enterprises. If, for a variety of reasons, the scheme is not adopted on pig farms, appropriate schemes with similar objectives must be evolved to recognise increasing levels of competence in stockpeople. This may take the form of 'in-house' schemes on large pig farms or within pig farming companies.

Within-farm career development schemes

The career development of stockpeople within farms could be catered for in a variety of ways. For example, Stockmanship Grade Levels 1 (Basic level) to 5 (a very high level of competence including man management abilities) should be defined as objectively as possible (see Table 8.5). Stockpeople would move up the scale on the strength of their performance based on a combination of developing an increasingly larger range of skills to a specified high level of competence and on the evaluation of their increased understanding of the pig and its requirements. This would help to create a more progressive career structure within pig farms or pig farming companies and the recognition of higher levels of competence and attainment through a combination of paper qualifications, increased

Table 8.5 Possible basis of a within-farm (company) career development plan for stockpeople

| Current stockperson category | Target career grade | Basis of evaluation | | | Rewards |
		Understanding of the pig	Pig handling skills	Man management	
New entrant to pig industry	1	✓	✓		Attainment certificate, salary and/or other rewards
Grade 1	2	✓	✓		as above
Grade 2	3	✓	✓	✓	as above
Grade 3	4	✓	✓	✓	as above
Grade 4	5	✓	✓	✓	as above

salary and other forms of award. Such an arrangement would provide a basis for progressive training and recognition of increasing levels of understanding and competence following such training.

Summary

A summary of the ways in which training in stockmanship can benefit both the stockperson and the employer is presented in Figures 8.1 and 8.2.

Provision of adequate training for stockpeople is important not only in relation to their competence but also to their job satisfaction. The factors, including effective training, which contribute to job satisfaction are dealt with in the next chapter.

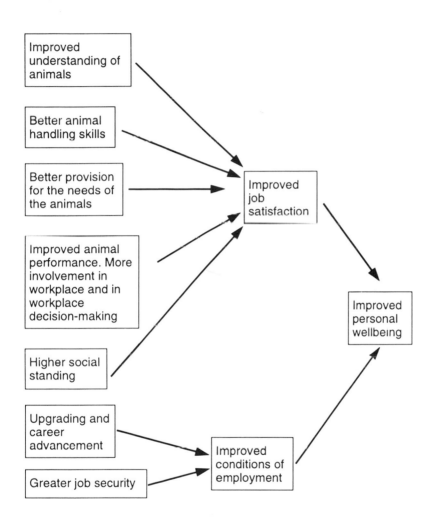

Figure 8.1   Benefits which a stockperson can derive from training

117

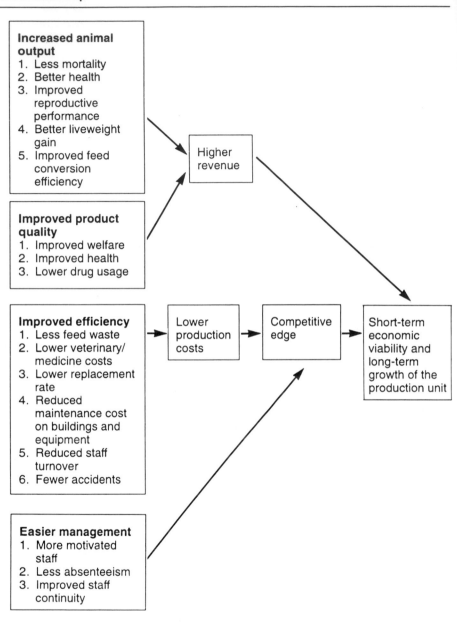

Figure 8.2   Benefits associated with the training of stockpeople from the employer's viewpoint

9

Job satisfaction and dissatisfaction in stockpeople

The pervasiveness of good job satisfaction in individual stockpeople and of a good team spirit in a workplace will make the work light, enjoyable and fulfilling. The pigs, the employees and the employer will all be beneficiaries.

The process, outlined in Chapter 6, of attracting more suitable applicants to a stockperson post and then selecting, as objectively as possible, the one with the best attributes for the job is a good starting point. The suitable employee selected will have a strong desire to be successful in looking after the stock for which he/she has been made responsible, and in improving the stock's wellbeing, health and performance. With such aspirations, it is important also that the employee can clearly anticipate recognition for such attainments in the form of career advancement along with cash and non-cash rewards. Whether or not these most desirable objectives are attained will be dependent on the owner/manager of the pig farm having similar objectives, and therefore providing the conditions necessary to maintain this initial motivation at a high level and at the same time creating the circumstances conducive to a continued high level of job satisfaction.

Any job carries with it some aspects which give satisfaction to the job holder and others which are a source of dissatisfaction. These are the 'good points' and 'bad points' felt about a job. Some of these points are transitory and soon forgotten while others are lasting or recurring and have an impact on personal behaviour and actions.

It can be expected that a stockman will take into account the amount of job satisfaction and dissatisfaction being felt when making the big decisions about work. 'Do I give everything to this job? Do I stay in the job? Should I accept this alternative job offer?' Also the employer would be wise to recognise that some factors create job satisfaction while others are conducive to job dissatisfaction for the stockman. This presents a management opportunity for even the most able and enlightened employer to find ways of further increasing job satisfaction and decreasing job dissatisfaction.

In this chapter these aspects relating to the good and bad points of a job are discussed from the point of view of both the stockman as employee and the owner/manager as employer. The essential components of motivation will be dealt with in the following chapter. At this stage, job satisfaction and the major factors affecting it will be discussed.

Job Satisfaction

Locke defined job satisfaction as 'a pleasurable or positive emotional state resulting from the appraisal of one's job or job experience.' It is a positive emotional response to a job situation. Thus, monetary gain is only one of the rewards a worker derives from his job.

The stockman, working in an animal production enterprise, can be looked upon as a seller; he sells his labour (time, skills and personal qualities) and in return wants to get:

- a high price for what he has to sell (good wages and perquisites)
- the maximum net satisfaction from the job (personal welfare in addition to material rewards)

He may have to trade-off somewhat between price and job satisfaction (see Figure 9.1).

In this example, the job offering the highest price to the stockman for his labour is also the one which provides least job satisfaction; this may, for example, be a post on a farm where processing, catering and hotel waste is fed; feeding is messy, pig mortality is high, feed costs are low and high wages have to be paid to attract labour. The job offering the lowest wage, on the other hand, may be on a traditionally managed family farm which is unable to take advantage of the economies of scale. However, the pigs are kept comfortably, mortality is low, breeding performance is good and the job provides much variety, including work in the farrowing, breeding and nursery sections.

Of course, the stockperson may find that there is little difference between the wages offered by various employers. For example, in Britain the Government-established Agriculture Wages Board for Scotland and another for England and Wales provide some control

Figure 9.1   Example of a situation where there is a negative correlation between the level of job satisfaction and wages on different pig farms

over wages. These Boards fix a minimum wage rate and a maximum value for certain perquisites such as housing. It is possible to find many employers offering these minimum wage rates. Faced with a choice of employer, all offering similar wages and perquisites, the stockman can then try to choose his employer on the basis of the amount of net satisfaction he is likely to derive from the job.

This job net satisfaction is the difference between job satisfaction and dissatisfaction.

Net satisfaction of the job	=	(amount of job satisfaction)	−	(amount of job dissatisfaction)

IMPORTANCE OF JOB SATISFACTION TO THE EMPLOYEE

Job satisfaction is the personal wellbeing (or in other words, the positive emotional feeling) obtained from the work situation. When deriving great job satisfaction, the stockman goes home, speaks kindly to and strokes the cat.

Job dissatisfaction is the personal ill feeling (or the negative emotional feeling) obtained from the work situation. When very dissatisfied with his job, the stockman goes home, shouts at and kicks the cat!

Thus the level of job satisfaction of the stockperson influences his behaviour both at work and outside the work situation.

IMPORTANCE OF JOB SATISFACTION TO THE EMPLOYER

The employer of stockpersons is a buyer; he buys labour and in return wants to get:

- adequate income (or profit)
- well cared for animals

Over time, a farm business must make a profit or it will fail in a competitive market economy. The employer has to evaluate the amount of benefit to the farm from employing the stockman and the cost to the farm of employing him.

The benefits to the farm which can be attributed to the stockman will depend on:

- the quantity and quality of the output in relation to the inputs
- the value of the output as indicated by the market price in relation to the cost of inputs

On the cost side the main determinants are:

- wages
- any perquisites
- government tax on jobs (for example, employers' national insurance contribution)

But there can be additional costs associated with:

- lateness and poor time keeping
- absenteeism
- labour turnover (cost of attracting, recruiting, selecting, inducting and training a replacement stockman in lieu of staff leaving for reasons other than retirement)
- stock mortality, poor reproductive and growth performance
- materials (such as feed) wastage

(Left)   *When deriving great job satisfaction, the stockman goes home and speaks kindly to the cat and strokes it.*
(Below)   *When very dissatisfied with his job, the stockman goes home, shouts at and kicks the cat.*

Vroom (1964) found significant correlations between job satisfaction and staff turnover, absenteeism and even better mental and physical health. Employees with higher levels of job satisfaction also had fewer on-the-job accidents. Thus, the employer's control of the factors affecting the state of job satisfaction and dissatisfaction for the stockman has a direct connection with factors affecting animal welfare, health, performance and profit.

JOB SATISFACTION IN RELATION TO PERSONAL ACHIEVEMENT NEEDS

The priorities an employee places on factors associated with a job are dependent on current levels of provision of personal achievement needs.

In a major study of human needs, the eminent American psychologist Maslow identified a scale for the individual as follows:

Level 1. Basic survival needs (food, clothing and shelter)
Level 2. Security needs
Level 3. Social belonging and acceptance by peers
Level 4. Social contribution and responsibility leading to increased esteem and status
Level 5. Full development of one's potential

This hierarchy of human needs is depicted in Figure 9.2.

At the base level are the basic survival needs such as food, clothing, warmth and shelter. In developed countries most full-time permanent employees in agriculture should have satisfied these needs. This level of need having been achieved, each worker will next strive for his and his family's security, not only in the short term but also for the long term. Satisfying this need is likely to become a higher priority in times of increased economic uncertainty.

Once survival needs and those for security are satisfied, the worker's priority will then be to feel accepted by his peers. It is likely that an employee who does not feel accepted in this way will grow dissatisfied and seek another post. On the other hand, if he is readily accepted by his peers, he will then strive to achieve the next rung in the ladder, that is, a position of status and esteem. This involves the need for his worth at work and in the community to be fully recognised by his peers and by the general public. At the same time he wants to feel part of the business he has served to date and, to achieve the highest rung on the ladder, he seeks every opportunity to express his full potential in contributing to the development and

124

MASLOW'S HIERARCHY OF HUMAN NEEDS

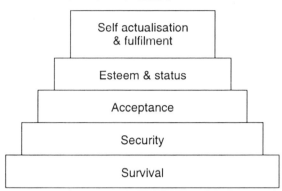

STOCKPERSON'S HIERARCHY OF NEEDS
(with suggested means of meeting these needs)

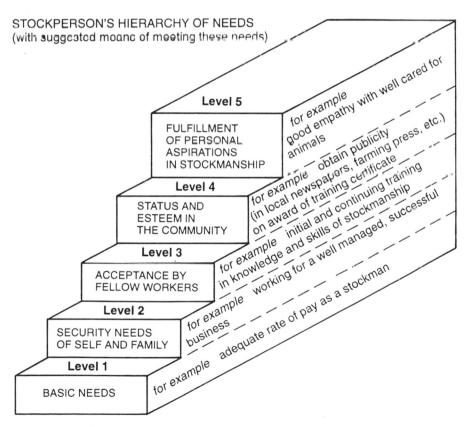

Figure 9.2 Stockmanship considered in terms of human needs

125

increasing success of the business. The two highest rungs on the ladder, which involve esteem and status along with self actualisation (expression of potential) and a sense of fulfilment, are usually referred to as 'higher level' needs. It is thus obvious that an employee's behaviour and immediate aspirations are very much influenced by the next level of need he has to satisfy.

Employers can influence the attainment of 'higher level' needs. Howard and his co-workers in Canada found that recognition of the high achievement of employees in forms ranging from bonuses to industry-wide awards can satisfy the need for esteem and status. Attainment of a responsible position in the farm structure that helps to develop an employee's capabilities to the full contributes to satisfying the highest order need, that is, for self actualisation and fulfilment.

It may well be that all workers, for a variety of reasons, will not aspire to the 'higher level' needs. The manual of the Agricultural Training Board in Britain on *Motivating Staff* (1983) suggested that employees have different degrees of personal achievement needs. Therefore, it is important that the achievement needs of each individual are assessed carefully and that full opportunities are provided to facilitate the attainment of personal goals. In this way, job satisfaction will be achieved and the usefulness of each employee to the business will be optimised.

The various factors which contribute to job satisfaction, including those highlighted in the Canadian study cited earlier, will now be discussed in further detail. Each factor can be viewed against the background of Maslow's scale of needs.

Factors contributing to job satisfaction

Herzberg proposed that job satisfaction is a function of achievement, recognition (cash and non cash rewards), the work itself and growth of the business. The major factors contributing to job satisfaction will now be outlined.

1. REWARD SYSTEMS

Reward systems embrace both cash (wages paid, bonuses etc.) and non-cash benefits. The relative importance attached by employees to such awards depends on prevailing circumstances. According to Howard and associates in Canada, money rewards are more attractive in periods of high inflation while in periods of high unemployment, job security becomes more important. The importance of wages may become less as an employee moves from the early-mid career phase into the later phase when status may assume higher priority. In this latter context, reward systems which appeal to employees' higher level needs, for example esteem and status, have been found to enhance employee productivity and to lower employee turnover (Howard et al. 1990). Rosen also found that rewards in the form of shares in the business increased the employees' pride in the work and interest in the long term future of the company.

Regarding bonus payment schemes based on performance, Howard and his associates (1990) reported that there is no consensus on their value: 'Such systems have advantages and disadvantages, they are highly situational and can even be counter-productive. However, in many situations, especially where the work is repetitive and boring, payment on the basis of performance can be advantageous.' The above research workers point out that if wages and other rewards are tied to performance attained, then management is indicating that management is in total control and is not appealing sufficiently to the basic desire in most workers to perform their duties effectively. Thus, the inherent motivation to work effectively for the good of the business may be reduced and consequently an employee's overall motivation and performance may decline.

A 'payment on results' approach can also be counter-productive when circumstances outside the control of employees cause poor pig performance. This may be a serious outbreak of a disease which has spread unexpectedly from another source or a sudden spell of excessively cold or excessively hot weather for which the farm has not adequately prepared. While it may be slightly easier to get a system of 'payment on results' to work effectively on farms with one employee, it is very difficult to get a system operating on larger farms which is fair to all employees. The good results achieved may be largely due to 3 to 4 key workers, yet it is extremely difficult to quantify accurately their positive effects on farm performance relative to a few careless and ineffective workers whose net contribution to an improvement in herd performance may well be a negative one. There is also a problem of fitting a new recruit into a 'payment on results' system for a team of stockpeople.

Incentive payment schemes can be based on results achieved by an individual employee, a group, a department or by the whole workforce. The owners/managers of pig farms have not been at all imaginative in recognising the efforts of individual staff, departments or of the entire workforce on a farm. Such inexpensive approaches as organising an annual meal and social evening for the staff and their spouses should be more prevalent than at present. At such an event, certificates could be awarded for special achievements, for example

- long service
- good empathy with the stock
- high piglet survival
- good conception rate
- maintenance of a high level of hygiene and tidiness in the work place
- most improved stockperson of the year
- most influential morale booster in the team
- the employee having the highest score in training course tests

The reason for the awards could be varied somewhat each year to allow for particular achievements in the year to be recognised or to acknowledge those new entrants who have contributed substantially to the team spirit.

Apart from the certificates awarded for particular characteristics and achievements, there could be one or two special prizes for an outstanding contribution to the performance of the team. In the case of a relatively large firm, management could arrange a peer nomination of the outstanding stockperson of the year based on a secret ballot. There could be a meal for a stockperson and spouse in a high class restaurant and/or a short holiday (e.g. long weekend) for two in an attractive resort.

Most employers in the pig industry fully appreciate the problems associated with a 'payment on results' system and prefer to offer good wages, the level of which will vary according to the complexity of the job and responsibility carried by the employee.

In the study of Howard and associates on pig farms in Ontario, employees stated that 'good wages' was the most important factor in job satisfaction. However, their acceptance of low wages (relative to other industries) indicates that factors other than financial compensation are influential in keeping them in their present job. These factors may include desire to work with farm animals, to live and work in the country and to be a member of a fairly small working team.

Related to reward systems are the aspects of working conditions which include working hours, weekend working practices, work load, days off, variety in work and other considerations in relation to

THE HOUSE OF GENETICS

National Pig Development Company

CERTIFICATE IN PIG HUSBANDRY

'Care of the farrowing sow and her litter'

This is to certify that   Audrey Wight   *has*

attended and completed the courses in the above subject and fulfilled the

necessary examination requirements.

Result: Distinction

Examiner ....
                            Signature

NPD Production Director ....
                            Signature

Certificate of attainment awarded to a senior stockperson employed by the National Pig Development Company in Britain following an in-house training course.

the range of jobs which have to be covered—from the mundane to the interesting and challenging. It is useful to formulate a policy on these aspects after having a full open discussion with the team of workers, in which the needs of individual staff for more or less work, attitudes to overtime and weekend working and the ideas of all are carefully noted.

All staff will be realistic in accepting that pigs have to be looked after for 7 days per week, that all tasks associated with tending the pigs, from the mundane to the attractive, have got to be done and that sometimes there is going to be a need to work extra hours. Having canvassed the views of staff, management is in a better position to draw up a plan of working practices which cover all the needs. Having been fully consulted on the issues and contributed substantially to the final plan, all staff will feel an obligation to support it fully and to make it work. This is likely to have an important effect on the efficient working of the team and the job satisfaction of its individual members.

These considerations are relevant to the concepts of job design and job enrichment.

2. JOB DESIGN AND JOB ENRICHMENT

Specialisation and division of labour whereby work is divided into small units, each of which is performed repetitively by an individual worker, usually leads to boredom and alienation which in turn causes job satisfaction and efficiency to fall. Job design is the process of deciding which tasks and responsibilities shall be undertaken by particular employees and the methods, systems and procedures for completing work. It concerns patterns of authority, responsibility, motivational demands and the relationship between the people involved. In order to avoid the negative effects of repetitive jobs on stockpeople, everyday jobs can be 're-designed' (job enrichment) so as to make them more interesting and varied.

To improve the acceptance of the 'low motivational jobs' of the general stockpeople (grades 1 and 2), job rotation within or even between farms could be tested, but the personal feelings of the actual stockpeople on the matter must be considered.

To 'enrich' the job of more experienced stockpeople, involving them in decision making is probably the best approach, since this stimulates their sense of participation and concern for the achievement of objectives.

3. ANIMAL COMFORT AND HEALTH

To a varying degree, all stockmen develop a relationship with the animals in their care. Seabrook (1982) states that in his work with animals 'man may assume one or other of the following roles: Boss animal, Mother substitute, Leader, Friend.' Even if it is difficult to provide supporting scientific evidence, few would deny that such factors as the health, body condition, appearance and comfort of livestock have a very important influence on the job satisfaction of the stockperson. In a study of stockpeople in Scotland, over a quarter were dissatisfied with the health and comfort of the pigs in their care (Segundo 1989).

The empathy which the great majority of dedicated stockpeople feel for their animals is such that these people will not be fully satisfied in their work until their charges are provided with conditions which are conducive to a high health status, a high degree of comfort and provided with all other (including nutritional) needs to ensure a good level of harmony with their environment.

4. THE WORKING ENVIRONMENT

This incorporates the physical space and facilities within the work place. It includes the housing for the pigs, equipment necessary for the job, the environment shared by the pigs and the stockpeople and the facilities available for the staff in terms of toilets, provisions for washing and showering, kitchen (cooking) and eating premises, and provision at 'break' times to attend to personal or family matters (e.g.

Facilities for staff at the entry point to the Sanfandila pig farm in Mexico are in the building to the left. They include showers, changing areas, washing facilities for clothes and a cooking/drinking area. There are identical separate facilities for men and women. The farm office is on the far right.

payphone) or recreational interests (e.g. simple 'sports' such as card games, dominoes and darts).

This category does not include the animals (dealt with under 'Animal comfort and health') and the people (dealt with under 'Within-farm interpersonal relations').

The effects of the working environment on the morale and the job satisfaction of pig stockpeople can be divided into:

- effects on physical health
- psychological effects

Studies in the USA, Sweden, Britain and the Netherlands have shown that workers in enclosed piggeries have high levels of respiratory injury, apparently caused or exacerbated by inhaled substances in their working environment. Three types of airborne hazards are known in piggeries, these being gases, dust and infectious agents such as bacteria. (Ragner Rylander, 1986; Watson, 1987)

In the study by Segundo and his colleagues on factors influencing the job satisfaction of pig stockpeople in Scotland, several components of the working environment caused concern. One third felt dissatisfied with the quality of the physical environment provided,

Maintenance of high standards of cleanliness and hygiene in the workplace is vital to human and animal health, job satisfaction and product quality. (Courtesy of *Dairy Farmer*)

the main focus being high dust levels. Other aspects over which there were lesser degrees of dissatisfaction were the lack of provision of 1) kitchen facilities, 2) showers and 3) protective clothing; 53%, 60% and 66% of farms respectively did not provide such facilities. It is surprising that more workers did not complain about failure to make provision for these basic facilities. This may be related partly to the fact that, traditionally, these facilities were not provided and partly to the tendency for farm workers in general not to complain about their job or job related factors but to get on with their work and 'make the best of things'. However, it would be most unwise on the part of employers to take advantage of this relative contentedness of stockpeople in their work. In general, stockpeople do an excellent job and they deserve a better working environment than many provide. Moreover, if recruits of high calibre are to be attracted to the industry and good stockpeople are to be retained in employment for a lengthy period, then it is imperative that all employers pay a great deal more attention than hitherto to the working environment and provisions within the workplace which are accepted as the norm in other industries.

5. WITHIN-FARM INTERPERSONAL RELATIONS

Another important factor affecting the job satisfaction of stockpeople is their relationship with colleagues which, of course, includes their superiors and subordinates. Armstrong and Lloyd (1983) made the following comment on the subject. 'A relationship, in a broad sense, is the position of one person to another. It extends to more than the formal relative position of superior, subordinate or peer. It is a function of all the circumstances surrounding the individuals i.e. their perception, values, expectations and the personal characteristics which each brings to the situation; their discrete properties which may be friendly or hostile, dominant or dependent.'

A successful relationship may be said to be one in which a state of equilibrium exists deriving from mutual modifications of behaviour —'give and take'.

The larger the group of workers employed in a business the more difficult it is to get all working as a completely harmonious team. 'Two's company and three's a crowd' is an old saying which sums up the difficulty of achieving harmonious working in a larger group. With a workforce of only 2 the individuals have merely to get their relationship with each other worked out. With 3 workers, A, B and C, there are 3 sets of relationships namely AB, AC and BC to sort out before they can become an effective team. With 4 and 5 workers there will be 6 and 10 individual person to person relationships to consider and with even greater numbers of employees the number of lines of communication between individuals multiply geometrically. Thus, relationships within a group of people are correspondingly more complex than those between just 2. When a working group is first formed there is a natural initial striving for dominance on the part of most employees, but once the 'peck order' has been settled by a combination of organisational structure and an acceptance of one's place in the hierarchy after initial sparring for position, the group can become an effective working team. However, each additional member to a group creates a new dimension of actions and reactions, threatens the established peck order, and the group acquires new characteristics.

Any disturbance to the peck order creates problems, and the sense of purpose and performance of the group may be temporarily diminished in the preoccupation with personality adjustments. Michael Argyle in his book *The Psychology of Interpersonal Behaviour* describes the stages of group development as being:

- forming
- storming
- norming
- performing

These 4 stages in the 'evolution' of an effective working group involve: 1) testing each other out, 2) conflict between pairs of individuals, 3) establishment of a stable hierarchy or chain of command and 4) working together effectively as a team.

Michael Argyle continues 'To some degree, each new member throws the group back to stage one and the whole process starts again. . . . Harmonious relationships are the best insulation against breakdowns in relations or organisations. Management has an important part to play in sustaining such relationships. It is said that management gets the work attitudes it deserves—the situation of action and reaction already referred to.

For example:

- criticise the man (as opposed to the work) and you get a self justification response
- be unfair and you get resentment
- pressurise and you get resistance
- plan badly and you get frustration

134

- be inconsistent and you lose credibility
- take all the decisions yourself and you get disinterest and non-commitment

Such a list is endless and one is forced to the inevitable conclusion that the causes of poor relationships between individuals on a pig farm are generally within management's control.'

In the study by Segundo and his colleagues on pig farms in Scotland, interpersonal relations were considered significantly more important than all other factors. Although the existence of good relations within the farm staff appears to be one of the main sources of job satisfaction, attitudes of other workers such as carelessness, lack of willingness to cooperate with others and lack of commitment to the work caused annoyance and greatly lowered the job satisfaction of one quarter of the employees involved in the study by Segundo et al. (1989).

In view of the crucial importance of good relations among all farm staff to job satisfaction, morale and their efficiency as a team in influencing the wellbeing and productivity of the pigs in their care, every possible effort must be made by management to recruit personnel with the appropriate attitude, interests and skills, to provide good working conditions and to do everything possible to develop and maintain an effective working team. The pervasiveness of a good team spirit in a workplace 'will make the work light', enjoyable and fulfilling. The pigs, the employees and the employer will all be beneficiaries.

6. ATTAINMENT OF FARM'S PRODUCTION TARGETS

When new recruits with the appropriate attitude and interest are appointed to pig stockperson posts they will be ambitious to be successful in their work. Such aspirations must be matched by those of the owner/manager of the business. This scenario was described by Cleary in Australia in the following terms:

'Good employees want to be successful and to believe that their efforts can improve the quality, health and output of the piggery in which they work. Such people need to perceive that their positions offer career growth and that the piggery owners are equally keen for the units to thrive.'

Although stockpeople carry less responsibility for farm performance than managers, it was obvious from the study of Segundo and associates in Scotland that employees on pig farms considered that farm performance was important to them and attainment of good levels of production was an important source of job satisfaction.

It is most important that managers are fully aware of this attitude in dedicated staff and play their full part in providing stockpeople with the appropriate pigs (good genotype and high health), buildings, space, climatic environment, equipment, working conditions, new colleagues of good calibre, continuing education/training and management style to make sure that these high aspirations can achieve the fulfilment that they deserve in the operation of the business. Such adequate provision of resources by management is in the best interests of the pigs, employees and the business both in the short and long term.

7. TRAINING

Training and continuing education of stockpeople in both understanding of the pig and in the skills necessary in handling and tending its needs is such an important component influencing the quality of stockmanship that it has a chapter of its own. Yet in the study by Segundo and associates in Scotland, over one third of employees were dissatisfied with the amount of both on-farm and off-farm training received.

Cleary, in Australia, points the finger accusingly at the amount of training provided to new employees at the start of a hoped-for career in pig production. He finds that poor induction to a new job as a pig stockperson (relative to good induction procedures which should include training in the understanding of the pigs, their needs and in handling skills) leads to discontent, poor quality work, low productivity and high labour turnover. It is no coincidence therefore that labour turnover on pig farms is highest among those employees with only a few months' service. Cleary found that well planned training improved both the capabilities and job satisfaction of stockpeople. In addition to the usefulness of more formal training, Cleary makes the following observations. 'Trade nights which provide information on commercial products and discussion group meetings and seminars which inform people about research are other means of providing informal training. Our own experience suggests that piggery owners who encourage their staff to attend such functions are rewarded with lower staff turnover rates, higher productivity and improved job satisfaction.'

Thus, a range of approaches to improve the understanding, awareness and skills of pig stockpeople have important influences on job satisfaction.

8. THE IMAGE OF THE JOB

Beynon (1991) has drawn attention to the image of the job of pig stockperson. In response to a questionnaire prior to a training course in England, the stockpeople indicated that they were reasonably happy with their job. However, their families were less enthusiastic and regarded working with pigs as having a low status. Beynon comments 'The public image of the swineherd continues.' Answers to further questions indicated that a high proportion of stockpersons who were contented with their work would not recommend the job to a young person and certainly not to their own family. This is a disturbing finding and indicates that a great deal of thought must be given to such aspects as job titles, career structure and working conditions.

Some fairly simple steps may be reasonably effective in changing the image of the stockperson's job in the eyes of his family and the community. Regarding job titles, the term 'stockman' has a more pleasant ring to it than 'pigman', while 'livestock technician' might to many be an even more acceptable title. With an effective career structure as suggested in Chapter 8 in operation, the 'Livestock Technician Grade 1' could then be promoted to Technician Grades 2, 3, 4 and 5 on the basis of experience, training, understanding, skills and achievements in terms of pig health, performance and production efficiency as well as on man management abilities at the Grade 3 to 5 levels.

Examples of improved working conditions were cited earlier but the provision by the company of clean protective clothing combined with washing and showering facilities would at least ensure that the stockperson left the dust and odours associated with pigs behind at the place of work instead of being accompanied home by them to annoy the family and pollute the household environment.

9. MANAGEMENT STYLE

All of the foregoing factors which impinge on the job satisfaction of the pig stockperson are controlled, or at least can be influenced, by management. Therefore, while it is convenient to consider the role of management style last, it has the most important influence on the contentedness of the stockperson with his job.

On the topic of management or leadership, Lloyd (1982) made the following relevant statement: 'Leadership is the true core mechanism by which management intentions are transferred into business activity. It works through people; it is about people.' Three main styles of leadership in the farm business can be recognised:

- *Autocratic leadership*. This assumes that the boss knows best and that an employee will stay in his job and work hard because he has a lot to lose and because it is expected of him. It is a legacy of the days when there was a strict social division between order givers and order takers and when the sack was a powerful sanction.
- *Laissez faire leadership*. This is the opposite of autocracy. Such leaders believe in non-interference. This type of leadership expects results from staff in situations where staff lack (1) the knowledge of the firm's expectations, (2) relevant information on which to make decisions (3) feedback and discussion of calculated performance trends and (4) even the necessary ability in management or husbandry to do the job.
- *Democratic leadership*. This style of leadership is not intermediate between the foregoing two styles. It makes very positive demands on the leader while liberating the abilities and application of the subordinate. Democratic leadership does not imply 'rule of the majority'. The leader's job is to manage using a participative style in which he inspires high aspirations in staff towards their work, keeps them fully informed in matters appropriate to their work and consults them on all matters relevant to it; it allows staff to take appropriate work place decisions, set work targets and monitor progress and results in consultation with him. The democratic leader is the type of leader most likely to achieve success in managing his staff and his business in today's pig farming situation.

The leadership style of many people is not exclusively of any one type but a combination; or perhaps the leader swings from one style to another at different times and with different people. Tannenbaun and Schmidt (1958) express the different leadership styles and their intermediate degrees in Figure 9.3.

Other aspects of the management role are those related to the quality and quantity of communication with staff. Kahan and Quinn (1969) drew attention to three common causes of stress at work, which are strongly related to poor communication between management and staff.

These are role ambiguity, role conflict and role overload. A 'role' here is defined as a set of behaviours expected of anyone occupying a particular job. Role ambiguity refers to situations where information

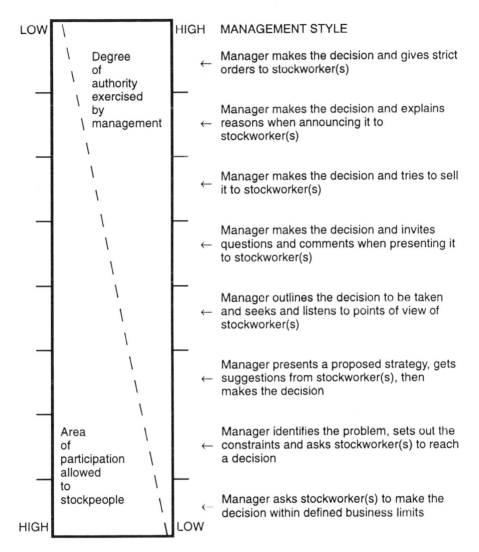

Figure 9.3   Examples of different management styles in animal production systems

available to the job holder about what is expected of him is inadequate. Role conflict is where two or more individuals expect different incompatible behaviours of the same individual. And role overload exists where a worker experiences a conflict of priorities because some expectations which people have of him can be met while others cannot.

In the study by Segundo and associates in Scotland, management style was found to be one of the largest sources of potential dissatisfaction of stockpeople. Generally speaking, more than a third of the stockpeople expressed dissatisfaction with the lack of communication between management and staff, lack of involvement and lack of meetings to discuss farm performance. When stockpeople were asked about the main source of dissatisfaction in their job, nearly a quarter of the replies were related to implied criticisms of management style.

Howard and his associates obtained similar findings from their studies on pig farms in Ontario, Canada. They reported as follows: 'Staff meetings, whether formal or informal were rarely held. The absence of staff meetings is an indicator of a lack of communication between employer and employee. Further evidence of a lack of communication was that few farms had performance appraisals. Forty seven per cent of the employees did not know their employer's goals (e.g. growth and expansion, debt reduction), which is not surprising given that 61% of the employers did not have written goals.'

The balance between job satisfaction and job dissatisfaction

Thus there are many factors which increase the satisfaction of stockpeople with their work and just as many that reduce such satisfaction. According to Herzberg, an employee will stay in a job as long as the satisfaction received is greater than the dissatisfaction caused by the job.

Howard and associates in their studies in Canada observed that 'as with any attitude, satisfaction cannot be observed, but must be assessed from either behaviour or statements.'

Accordingly, employers and employees on pig farms in Ontario were asked to rank the relative importance of 10 job factors which were considered to be relevant to the level of job satisfaction of employees. The results are summarised in Tables 9.1 and 9.2.

In deciding on their overall level of satisfaction or dissatisfaction with a job, employees are likely to weigh up all the influential factors in relation to their own particular aspirations. According to Howard and his associates, 'the satisfaction employees feel about a job is balanced against dissatisfaction caused by regulations, supervision,

Table 9.1   Employees' degree of satisfaction with selected job factors compared with employers' perception of employee satisfaction on pig farms in Canada

	Job factor	Employees % satisfied	Employers' judgement regarding % of satisfied employees
1.	Good working conditions	89.3	100
2.	Complete understanding of operation	95.9	95.3
3.	A feeling of being involved in operation	94.2	97.9
4.	Sympathetic help in problem solving	83.4	100
5.	Fair supervision	90	100
6.	Good wages	76	92.3
7.	Good job security	86.6	97.6
8.	Good opportunity for promotion	53.7	52.4
9.	Interesting work	91.7	92.9
10.	Companionship with other employees	85.9	83.4
		Ave. = 84.7	Ave. = 91.2

Source: Based on Howard, W. H., McEwan, K. A., Brinkman, G. L. and Christensen (1990)

Table 9.2   Comparison of rank order given by employees and employers to selected job factors on pig farms in Canada

Job factor	Employees' ranking of factors	Employers ranking of factors
Good wages	1	2
Good working conditions	2	1
Interesting work	3	3
A feeling of being involved	4	4
Complete understanding of operation	5	5
Good job security	6	6
Fair supervision	7	7
Good opportunity for promotion	8	10
Companionship with other employees	9	9
Sympathetic help with problems	10	8

Source: Based on Howard, W. H., McEwan, K. A., Brinkman, G. L. and Christensen (1990)

work conditions and (lack of) rewards.' They state that the balance between the 2 forces is delicate, but most important, since satisfaction is a key factor determining an employee's performance.

Summary

Figures 9.4 and 9.5 highlight the major factors contributing to job dissatisfaction and job satisfaction respectively while a summary of the main indicators of job dissatisfaction is presented in Figure 9.6.

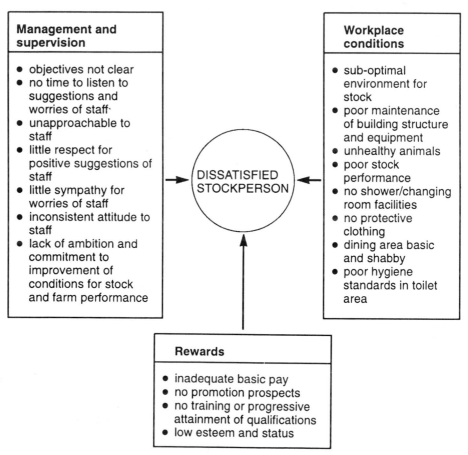

Management and supervision

- objectives not clear
- no time to listen to suggestions and worries of staff·
- unapproachable to staff
- little respect for positive suggestions of staff
- little sympathy for worries of staff
- inconsistent attitude to staff
- lack of ambition and commitment to improvement of conditions for stock and farm performance

DISSATISFIED STOCKPERSON

Workplace conditions

- sub-optimal environment for stock
- poor maintenance of building structure and equipment
- unhealthy animals
- poor stock performance
- no shower/changing room facilities
- no protective clothing
- dining area basic and shabby
- poor hygiene standards in toilet area

Rewards

- inadequate basic pay
- no promotion prospects
- no training or progressive attainment of qualifications
- low esteem and status

Figure 9.4   Major reasons for job dissatisfaction

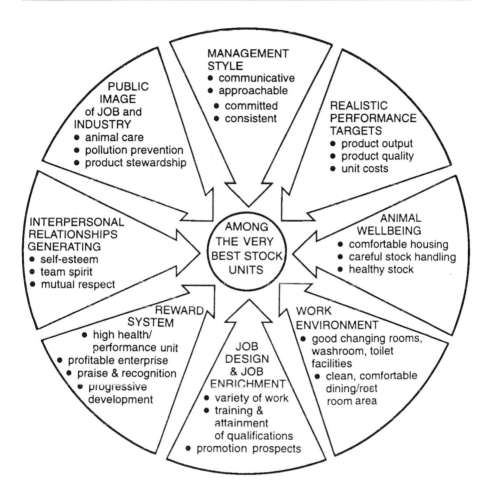

Figure 9.5   Approaches to achieving good job satisfaction

JOB SATISFACTION IN GENERAL

When potentially good employees are attracted to a stockperson's post and subsequently leave because of a lack of job satisfaction, such turnover of staff tends to be a 'disruptive, costly process'. Therefore every effort must be made, within the limits of available resources, to do everything possible to ensure a feeling of job satisfaction. The onus is on employers/managers to implement this awareness effectively in the management of the stockpeople at their disposal. When

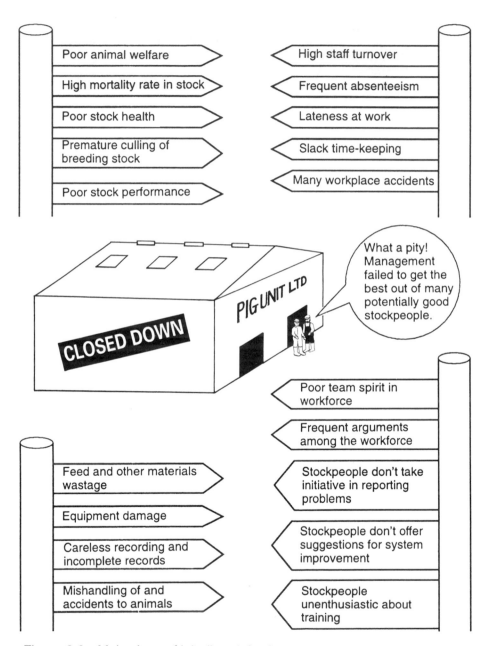

Figure 9.6   Main signs of job dissatisfaction

they achieve this objective the pressures on themselves as managers will be less since their stockpeople will then have a similar degree of motivation as themselves to make the business successful.

Other aspects related to motivation of staff will be outlined in the next chapter.

10

Motivation of stockpeople

For a worker or workforce to be productive, not only must work methods be effective, easy and safe, and the staff of the right calibre, but they must derive satisfaction from their work and be motivated towards its accomplishment at high standards.
(Lloyd 1975)

Stockpeople, in caring for their animals, carry out their duties more effectively if they are motivated. Before discussing factors involved it is useful to discuss the concept of motivation.

The concept of motivation

The word 'motivation' is derived from the term 'motive' which signifies a stimulus. All our actions and those of a stockperson are driven by motives or stimuli of one kind or another thus:

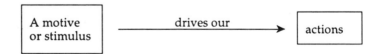

The action taken following a stimulus will in turn result in some kind of outcome or achievement. For example, a stockperson who is motivated to look after his animals with great care will improve their health, welfare and performance. These influences in turn will make the enterprise more profitable and the industry more successful. Such achievements will provide the stockman with rewards for his increased efforts in applying his skills thus:

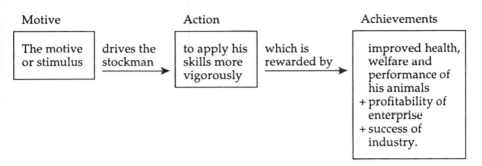

These high levels of achievement by the stockperson in turn provide him with a great deal of personal satisfaction thus:

All that we need now to complete the process is to insert a trigger or motivating factor which sets off the whole sequence of events, thus:

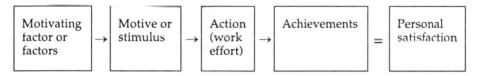

Success breeds success in the motivation of stockpeople and this can drive the stockperson on to progressively more effective efforts thus:

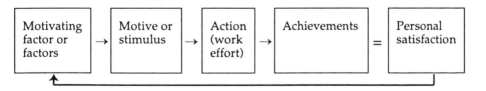

Since the motivating factors have such a crucial role to play in this most desirable sequence of events, it is important to be fully aware of motivating and negative influences on the stockperson.

Removing negative influences

The first essential is to remove factors which have a negative or 'demotivating' influence. As indicated in the last chapter, the factors which cause job dissatisfaction include:

- poor workplace conditions e.g. dusty premises, poor kitchen/ dining facilities, poor provisions for personal hygiene such as lack of showers, and failure to provide sufficiently high standards of accommodation for livestock
- oppressive control and supervision; this may involve autocratic management and poor communication between management and staff
- difficult relationships with fellow members of staff and poor team spirit
- repetitive nature of the work leading to boredom
- low pay
- unfair delegation of duties and workload by team leader/ manager
- excessive weekend work
- excessively long working hours

Removing such sources of job dissatisfaction was the subject of the last chapter. But the correction of these deficiencies and problems does not in itself motivate the workforce. According to Lloyd (1975), such correction 'merely produces a condition of no dissatisfaction in which the workforce is generally cooperative and does not seek to change employment actively.' The work force will not be stimulated to give of its best by such a policy but it is receptive to motivators which may then be employed effectively.

Motivation and worker performance

While it might be expected, other things being equal, that appropriate motivation would improve worker performance, Howard et al. (1990) point out that the concepts of 'motivation and performance are not synonymous. Motivation represents an employee's desire to perform, while performance is the extent to which an individual can success-fully accomplish a task or achieve a goal. Performance is also affected by ability and opportunity.' Facilities and training also affect perfor-mance. The relationship between motivation and performance can be described by the model shown in Figure 10.1.

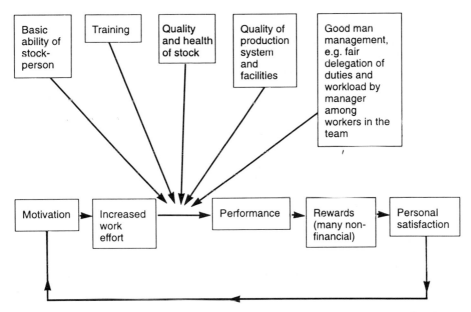

Figure 10.1   *Cyclical influences on motivation and other factors affecting performance*

The anticipated rewards (many being non-financial, such as personal satisfaction) trigger motivation to exert increased effort. The resulting performance will be a function not only of the motivation of the worker but of his ability and of the facilities/resources he has at his disposal. When performance is enhanced, the worker is motivated by the response (rewards) and enhanced degree of satisfaction to renewed efforts.

MOTIVATORS AND ATTITUDES TO WORK

Among the motivators which may set this continuing sequence of events in motion are greater recognition of worth, being held in higher esteem by peers and colleagues, greater involvement in decision making and challenge in the workplace and, in some circumstances, greater financial rewards.

For a worker or workforce to be productive, Lloyd (1975) points out that 'not only must work methods be effective, easy and safe, and the staff of the right calibre, but they must derive satisfaction from their work and be motivated towards its accomplishment at high standards.

Motivation derives from a desire to satisfy unrequited needs or avoid undesirable experiences. It determines attitudes and behaviour in both working and domestic environments. Many industrial research studies have shown that the attitudes of staff can greatly affect their performance at work and the reliability of their attendance at work.'

SATISFYING OF BASIC NEEDS AND FEELING OF INVOLVEMENT IN THE WORK PROCESS

In order to induce the correct attitude to work it is important to draw on the experiences from other industries and to consider the hierarchical order of human needs as discussed in Chapter 9, which dealt with job satisfaction. The basic need is for survival, followed by security. Once these needs are satisfied 'man derives considerable satisfaction from the status which a job allows him amongst his fellows both at work and outwith the work environment, and from the feeling of involvement in the work process itself and amongst work fellows, business associates and social contacts. Most workers also appreciate the opportunity of making full use of their varied talents and skills and when suitably trained they enjoy the opportunity of responsibility and authority which comes from the challenge offered by a job' (Lloyd 1975).

PAY AND CONDITIONS OF EMPLOYMENT

Pay is important to a worker's self-esteem as well as to his status and social standing. However, Lloyd makes the point that 'in a reasonably well paid job the most influential factors in the person's attitude and behaviour derive from the conditions of employment.'

STAFF MANAGEMENT POLICIES

This places considerable onus on management to provide the most stimulating conditions for staff and Lloyd (1975) described this responsibility of management as follows:

'Effective staff management today involves an understanding of staff needs, ambitions and dislikes and a willingness to adapt the policies and practices of the firm so as to make the best use of this knowledge. Sound management should ensure that:

- staff are equipped to cope with the demands and challenge of their work through selection, training and experience

- their work conditions are pleasant and safe and their pay sufficiently rewarding
- they are provided with sufficient tools, materials of the right quality, information and personal authority to use their skills to full effect

Management should also attempt to involve and stimulate staff to give its total effort in such a way that it derives satisfaction from the work.'

Lloyd, along with his colleague Armstrong at the University of Reading, based many of his concepts on those of an American industrial psychologist, Professor Likert, who contended that:

'Business success and staff productivity, satisfaction and wellbeing can be assured by a three-point formula which consists of:

a) promoting supportive relationships, i.e. relationships within a socially well-developed working group structure which the members see as being psychologically rewarding to them;
b) adopting group decision-making and supervision. This entails granting authority to working groups to take decisions regarding their work place activities and through the social pressures which arise within such a group to bring into line those members whose performance fails to conform to the group norm;
c) inspiring groups to set their own high working targets, i.e. encouraging staff to set their own work norms but at the same time communicating the management's high aspirations whereby the achievement needs of individuals may be aroused.'

IMPACT OF STAFF MANAGEMENT POLICIES

Through the investigation of industrial processes in America, Likert was able to show that businesses where the staff management policy conformed to this three-point formula exhibited high productivity, low labour turnover and high job satisfaction. He also demonstrated that where the staff management policies of a business were changed to accord with the formula, improved productivity, low labour turnover and easier supervision resulted.

Armstrong and Lloyd (1972) used the methods developed by Professor Likert to assess the adequacy of management staffs in agriculture in terms of such aspects as leadership, communication and interpersonal relationships.

They also accepted Likert's classification of the following four basic types of staff management:

1. exploitive authoritative
2. benevolent authoritative
3. consultative
4. participative group

Their later research in agricultural businesses eventually confirmed that as staff management style progressed from Type 1 to Type 4, so productivity, staff stability and profit improved.

CHARACTERISATION OF STAFF MANAGEMENT SYSTEM

The methods used by Armstrong and Lloyd to assess the quality of staff management systems in agricultural businesses are worthy of note.

They examined the following aspects of staff management systems.

- the leadership processes
- the motivational forces
- the communication processes
- the interpersonnel relationships
- the decision-making processes
- the target-setting processes
- the control processes
- performance-determining aspects of the work such as training and support facilities
- working conditions

Each of these 9 components was in turn subdivided into 4 to 6 aspects and these are detailed in Table 10.1.

Employees in the agricultural businesses were then asked to record:

- their personal experience of this factor in the business
- the personal experience they would like to have in relation to this factor

These records were then used to determine the adequacy of motivation provided for each of the components of staff management (see Figure 10.2).

It can be seen that the scores for the level of motivation provided by management were generally high with a mean score of 14.6 out of a maximum of 20. Scores for many of the components of management as assessed by staff were in excess of 15. Only 3 aspects of

Table 10.1   The Armstrong and Lloyd model for the description of the staff management quality of a business

Characteristic		Example aspects
Leadership processes	1	Interest shown by management in personal/family problems
	2	Confidence shown in subordinates' ability
	3	Understanding by management of work problems of subordinates
	4	Fairness of treatment between subordinates
	5	The degree of confidence which subordinates have in their superiors
Motivational forces	6	Appreciation of work performance shown by superior
	7	Extent of fear, cash rewards or personal satisfaction in determining the way staff work
	8	The extent to which subordinates' abilities are used
	9	The feeling of achievement derived from doing work well
	10	The extent to which staff are told how well or how badly they are doing
	11	How important the superior appears to regard the subordinate's work
Communication processes	12	How much subordinates are told about the objectives, aims, plans and performance of the business
	13	How much information passes between colleagues
	14	Attention given to complaints, comments, etc.
Relationships	15	Approachability of superior
	16	Degree of co-operation in the organisation
	17	Attitudes to colleagues
	18	Helpfulness of superiors and colleagues
	19	Style of treatment when mistakes are made
Decision-making processes	20	Degree of involvement in decisions
	21	Awareness of superiors of work problems when making decisions
	22	Degree of consultation concerning workplace decisions
Work targets	23	Awareness of the work goals of the business
	24	Degree of involvement of staff in setting work goals
	25	How far goals stretch the capabilities of staff
	26	Whether staff approve of the goals set
Control processes	27	Freedom to regulate speed of work
	28	Ability to influence others whose work affects their own
	29	Frequency of reviews of work progress
	30	Occurrence of conflicting instructions
Performance-determining factors	31	Adequacy of the methods or equipment
	32	Amount of material wasted
	33	Amount of time wasted
	34	Need for skilled instruction
Workplace conditions	35	Satisfactoriness of working hours, work premises, methods and content

Source: D. H. Lloyd, *Effective Staff Management*

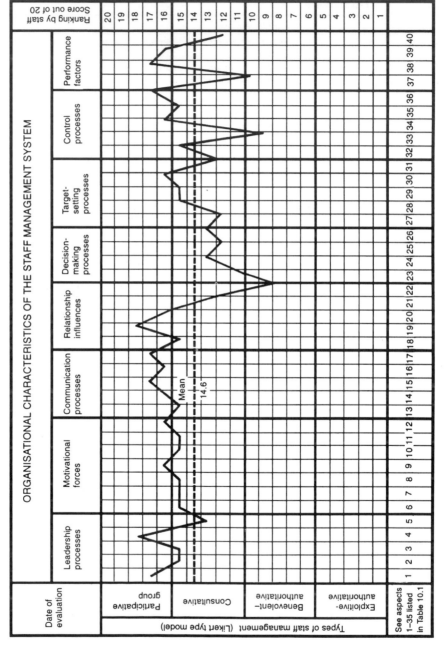

Figure 10.2   Staff management audit or evaluation profile (from Armstrong and Lloyd, 1972)

154

management failed to attain the median point of 10 in the scoring system.

These were:

Component

19   Style of treatment when mistakes are made (Score 7.5)

28   Ability to influence others whose work affects their own (Score 8.5)

31   Adequacy of the methods or equipment (Score 9.5)

This evaluation was termed a 'staff management evaluation profile' or a 'staff management audit' and proved effective in detecting defects in staff management procedures. The next obvious step is to set out to correct the deficiencies detected in components 19, 28 and 31.

Lloyd and Armstrong (1973) reported on the application of the staff management evaluation profile within a large dairy company with many herds and 60 employees, with a view to improving motivational forces. The company was already successful but among the steps taken following the audit were the following:

- traditional cash incentives were abandoned
- several factors causing job dissatisfaction were removed
- group decision making was promoted
- communications were improved in all directions between supervisors, manager and dairymen
- skills were improved through work related training
- motivation was increased through psychological rewards
- self-set production targets were formulated
- support facilities were improved

The result of the exercise was an increase in motivational level, reduced staff turnover and improved technical and economic viability.

PAY-FOR-PERFORMANCE REWARD SYSTEMS

Because pay is such a core part of the relationship between farmers (as employers) and stockmen (as employees), the way it is handled influences all other aspects of the relationship. This includes the motivation process.

Management has certain choices regarding the form of payment it adopts. The main forms of payment are:

Type A   Payment by timework
Type B   Payment by results
   ● piecework
   ● bonus
Type C   Payment by merit

The incentive schemes for payment by results can be based either on individual workers or on collections of workers (team group, section group, department group or whole farm).

The payment system adopted can affect the behaviour of management and workers. In Britain, the Agricultural Wages Board for England and Wales and the Scottish Agricultural Wages Board determine minimum wages for farm workers based on payment by timework. One can safely predict that a management which practises this form of payment scheme will, at least initially, be either dismissive of or unsympathetic towards the ideas on motivation under consideration in this chapter. However, management may choose a payment system with the intent of having a genuine motivational purpose. The recent trend in Britain, at least in non-agricultural sectors, is for firms to adopt and progress with schemes for payment by results and thus link work effort and achievement to pay rewards.

The studies of Howard and his colleagues on pig production enterprises in Canada indicate that systems for incentive payment schemes 'have both advantages and disadvantages, are highly situational and can even be counter-productive'. Mueller, Hollis, Johnson and Waldo (1983), on the basis of their experience of pig production units in the USA, specified basic principles for the operation of pay for performance reward systems. Based on these two studies, together with work done by the authors in Aberdeen, underlying principles and guidelines can be outlined for any payment by results scheme and thereafter the case for and against payment by results schemes can be made.

Basic guidelines for an effective payment by results scheme

● There must be a high degree of trust between employer/farmer and employee/stockman.
● There must be a broad area of common ground between the values held by the management and the personal values of the stockman.
● There must be a high degree of stability in the use of pay for performance. In other words, the pay for performance reward

system has to form a part of management's long-term strategy and not be used as a short-term tactic.

- The incentive should be based on performance levels that are observable, measurable and controllable by the employee.
- The earned incentive or bonus award should be paid promptly and regularly. The objective of the incentive or bonus is to motivate employees to perform above minimum levels. The rewards or lack of rewards where incentive levels are not met should reinforce this performance goal.
- The exact terms and conditions of incentive payments should be in writing and included in written employment agreements with a copy of the complete agreement provided to each employee affected by the arrangement.
- Base pay scales should be reasonable and adequate for services performed so that failure to achieve the incentive level of performance due to extraordinary causes does not result in a substandard pay level for the employee.
- The incentive plan should be reviewed regularly and modified as the employee gains in skills or is assigned to new duties and responsibilities.
- Each individual employee probably possesses a unique set of values, personal security goals, career objectives and long-term financial goals. It should be recognised that not all individuals will respond in the same way to a given incentive plan, nor should they be expected to respond uniformly.
- The incentive plan should not be a substitute for good labour relations, and in personal interactions between employers and employees, each employee should be treated with consideration and respect.

Changing from paying by timework (for instance £200 for a 42 hour working week) to paying by results is not a 'quick fix' solution to any sort of problem within the pig production unit. Neither will the adoption of a payment by results scheme result in the desired improved motivation towards the job without good groundwork preparation based on the above basic principles.

Some of the advantages and disadvantages of payment by results schemes are as follows:

Advantages

- The stockman is made more aware of the successes and failures within the business and is more involved in the challenge of pinpointing the reason (or reasons) for such successes and failures.

- There is an improvement in employee motivation and morale following a bonus or award for exceptional services, or the sharing with employees of part of a windfall success of the business.
- A carefully selected new entrant stockman is drawn more quickly and effectively into a group of dedicated and high quality stockpeople already operating as a very effective team on the farm because the existing workers have a strong vested interest in coaching and integrating the newcomer into the best ways of doing things.

Disadvantages

- The familiarity effect; for example, a traditional Christmas bonus can become expected and anticipated by the employee as part of his regular pay. In this case it no longer provides a motivational effect and is no longer evidently tied closely to productivity.
- The domino effect; for instance, bonus payments for farrowing unit staff based on numbers weaned can be upset by low numbers born due to the negligence of breeding section staff. Such a problem might be solved by paying the farrowing staff a bonus based on piglet survival percentage and weaning weight while initiating a bonus system for the breeding section staff based on farrowing rate and numbers born.
- The external circumstances effect; this disadvantage can arise in situations where productivity and profitability are influenced largely by extraneous factors. Employees may be on a good basic wage with an extra bonus payable on the basis of previously agreed indices of productivity and profitability. This is fine when results are good and employees receive a much needed extra boost to their finances at a time when spending is high such as at Christmas and holiday times. However, how disappointed will they feel after a bad year during which they may have worked even harder than before but when no bonus is payable? They may have achieved excellent results in physical terms (e.g. in weaners per litter, litters per sow per year and very high pig survival rates) but, because of very low pig prices, a financial loss has been incurred in the business and therefore no bonus is payable. On the other hand, productivity may be low for reasons outside the control of the workforce. There may have been a serious abortion storm in the pigs, dairy or suckler beef herd or sheep flock, the

cause of which had nothing to do with negligence on the part of the employees.

A good example of such unexpected problems experienced recently in Britain and parts of Europe has been the outbreak of Blue Ear disease. It has caused extremely heavy losses in the form of abortions and piglet mortality in herds where owners, managers and all employees appeared to have taken all possible precautions to prevent it. However, the spread of the causative virus via wind has been outside their control. When such a scourge is going through a herd, dedicated stockpeople are totally discouraged as they see sows aborting and piglets dying in great numbers and are unable to do very much to reduce the devastation. Such extreme disappointment is compounded by the thought that the prospect of their bonus on productivity and profitability, confidently anticipated before the disease outbreak, is now disappearing like the morning mist.

- The bickering team effect. For example, when a team of stockpeople are operating a large farrowing unit, most employees may be having a very favourable influence on the farrowing, the sows and the piglets, whereas the behaviour and working practices of one or two may be counter-productive. Since Hemsworth and his colleagues in Australia have found that the aversive practices of one stockperson can have the same effects in depressing pig welfare and performance as the good influences of 5 caring stockpeople, the net impact of the entire team on welfare and productivity may be nil and so no bonus on productivity is payable. This is very demotivating on those stockpeople who are having a very good influence on the wellbeing and productivity of the stock. Many will say that the obvious solution to this problem is to dismiss the stockperson displaying negligent and uncaring attitudes towards the stock, but such behaviour is often difficult to substantiate against possible subsequent claims for wrongful dismissal. The other approach is to attempt to correct and improve the methods of the wayward stockperson but such attempts again may not prove successful if the basic problem is the employee's lack of empathy towards his animals and of basic interest in the job. Such deficiencies, of course, should have been noted at the selection and recruitment stage and the person would not have been employed in the first place.

To sum up, there can be no general consensus on the desirability or otherwise of incentive payment schemes. It is up to each employer/manager to take stock of their own particular situation and to have in mind the characteristics and aspirations of their own employees as a basis for deciding whether bonuses based on higher performance attainments would, on balance, be helpful, or potentially counter-productive at times, in their specific business.

THE NEED FOR CONTINUOUS MOTIVATION

The need for motivation is continuous. As the stockperson responds to motivation by thinking how he can tend his animals more effectively and/or by increased work effort, he will usually be rewarded by the increased wellbeing, improved appearance and enhanced performance of the stock. This will encourage him to be progressively more effective. However, the motivators which set him going in the first place are likely to be less effective as time goes on in their impact on his thinking, his attitude and his effort (see the drawing on the facing page). Therefore, it is the responsibility of management to come up with new motivators over time. These may take a variety of forms such as financial rewards, increased responsibility, improved status in the workplace, delegation to represent the business at relevant conferences and discussion groups, participation in advanced training courses and increasing involvement in training of new and established employees on the farm.

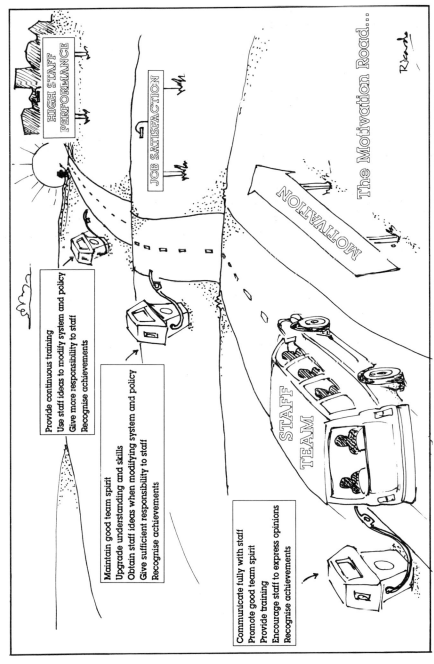

HIGH STAFF PERFORMANCE

JOB SATISFACTION

The Motivation Road....

MOTIVATION

STAFF TEAM

Provide continuous training
Use staff ideas to modify system and policy
Give more responsibility to staff
Recognise achievements

Maintain good team spirit
Upgrade understanding and skills
Obtain staff ideas when modifying system and policy
Give sufficient responsibility to staff
Recognise achievements

Communicate fully with staff
Promote good team spirit
Provide training
Encourage staff to express opinions
Recognise achievements

The need for motivation is continuous. It is the responsibility of management to 'refresh' staff by developing new initiatives to maintain motivation.

Summary

The main lessons to be gleaned from this chapter are outlined in Figures 10.3–10.5.

Motivation is a difficult concept for most people to fully appreciate. It has for a long time been an important subject for study by many eminent psychologists. They have still a great deal to learn about its origins and its impact. However, despite imperfect knowledge on the subject, we know that it has a powerful influence on the job satisfaction of the stockperson, on his attitudes to work, on his mental and physical efforts on behalf of the stock and the business and on the resulting influence on the health, welfare and performance of the animals in his care. In the pig production enterprise we must get rid of all the disrupting influences on the motivation of each stockperson and continually seek ways of motivating and remotivating each individual in the working team and the team as a whole.

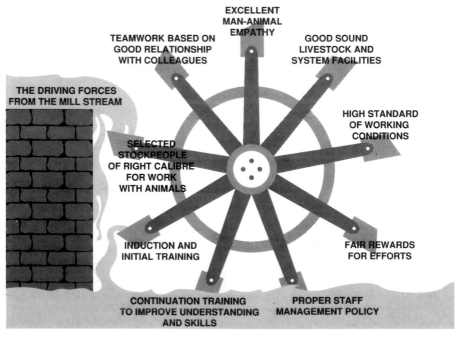

Figure 10.3   Main determinants of stockmanship

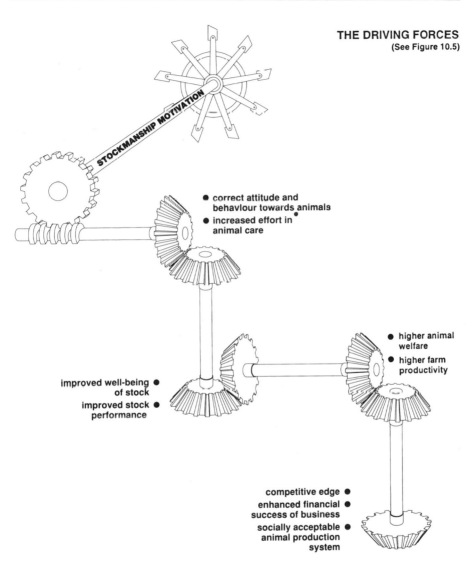

THE DRIVING FORCES
(See Figure 10.5)

- correct attitude and behaviour towards animals
- increased effort in animal care

- higher animal welfare
- higher farm productivity

improved well-being of stock
improved stock performance

competitive edge
enhanced financial success of business
socially acceptable animal production system

STOCKMANSHIP MOTIVATION

Figure 10.4   Effects of good stockmanship motivation

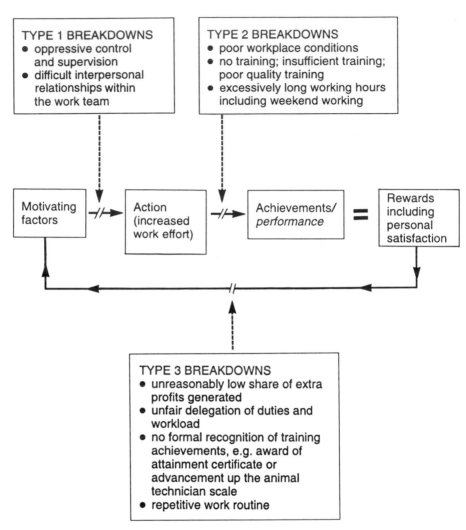

Figure 10.5   Causes of breakdown in the motivation machinery

11

Improving stockmanship—
the correct route to take

The more you study an important component of animal production systems such as stockmanship, the more clearly you see first, the weaknesses and thereafter, the most promising solutions in relation to improvement.

Current relevance of stockmanship

While stockmanship has always been of crucial significance in animal production, it has never been more important than at the present time. There are several reasons for its current significance in pig production:

- Its major influence on animal welfare
- Its marked influence on productivity and efficiency
- The apparent shortage of supply of suitable personnel willing to take up pig stockmanship as a career
- The increased economic pressures in pig production which demand higher output and efficiency
- The complexity of many of the jobs in pig production
- The large size of many pig enterprises necessitating teams of workers, a cooperative spirit and good teamwork of the full workforce.

In order to ensure the high quality of stockmanship which the pig, the pig producer and the pig industry require, the following provisions are necessary:

1. Making the job more attractive.
2. Objective selection of potential recruits on the basis of interest, attitude, understanding and ability. Empathy with animals is one important component.

165

3. Training to increase understanding and enhance technical skills.
4. Sufficient time to apply the enhanced skills and understanding to the improved care of the pig.
5. Motivation and re-motivation.
6. Job satisfaction which should follow if 1 to 5 are covered.
7. Retention in employment. If objectives 1 to 6 are achieved, labour turnover in the pig industry will be reduced. The longer a good, well-motivated stockperson is retained in employment, the more effective he will become.

Studies among stockpeople in Britain and other countries indicate the very pressing need to achieve objectives 1 to 7 if good stockmanship is to achieve the massive impact it can have on our pigs, our producers and our industry.

To date, the pig industry in Britain has largely resisted the temptation to reduce labour input and labour costs in an attempt to improve financial returns. Our labour input is considerably higher than that in the USA and Canada. It is no coincidence that the welfare and productivity of our pigs are also higher, on average, than in these countries.

To satisfy objectives 1 to 7, our labour costs, as a proportion of total costs, will undoubtedly increase. However, once these 7 objectives are achieved, our labour input will be more efficient and more cost-effective. For want of a suitable example from pig production, let us cite a relevant example from the broiler industry to try to support our claim.

Lloyd (1975) calculated alternative ways of increasing the profit margins for broilers by 30% (Table 11.1).

Table 11.1 Alternative strategies for increasing profit margins per bird in broilers by 30%

Strategy	+ 30% profit margin/bird
Birds per man	+90%
Viability	+1.67%
Food used	−2.5%

While economies in the deployment of labour must always be sought in pig production this should not be at the expense of care expended on the individual pig. Lloyd estimated that a very small (1.67%) increase in survival of broilers or a reduction in food requirement of 2.5%, achieved by more attention to stock and to the system, could each improve the profit margin by the equivalent of almost halving of the labour force (i.e. birds per man increased by 90%). Likewise on pig production enterprises throughout the world, great dividends can be achieved from investment in improved stockmanship.

Improving stockmanship

In setting out to improve stockmanship, the stockpeople them-selves, owners and managers of pig enterprises as well as the educators/trainers, researchers, the media and policy makers in any country can all be influential as indicated in Figure 11.1 (pages 168–169).

Improving the general level of stockmanship will not be easy but with soundly based, determined and sustained effort the potential rewards are enormous. Not only will the attainment of the goal demand action by all parties involved, but it will also demand attention at all stages, from the attraction of applicants through the selection, induction, training, rewarding and motivational processes. As research continues to identify and quantify the enormous benefits of good stockmanship, the need for action becomes ever more pressing.

The steps in the vital process from the attraction of inexperienced applicants for a stockperson post to the progressive development of the successful applicant into a knowledgeable, highly skilled and well-motivated stockperson with a very desirable influence on the animals in his care is summarised in Figure 11.2 (page 170).

Thus, the route from being a novice to being a highly skilled stockperson is long and painstaking. However, if all the simple principles are applied and the basic steps taken, it is a sure way to end up with the highly skilled and well-motivated people which we need to provide excellent care in our livestock production systems.

Stockmanship—an improved definition

When providing tentative definitions of stockmanship in an earlier chapter, we gave an undertaking to provide an improved definition after we had considered all relevant components and influences more fully. We have now reached this stage and our improved

(Text continues on page 171)

Figure 11.1   Collective efforts in improving stockmanship

What stockpeople should do	• Recognise that knowledge of the animal's needs is an essential requirement of competency in stockmanship • Accept that talking to and touching an animal are important aspects of good stockmanship • Strive for consistency in animal handling skills • Give sufficient thought to the abilities which make for excellence in stockmanship • Remind management when a lapse occurs in provision of continuing vocational training • Collaborate with management on reward pay system linked to improvements attained in stockmanship • Continuous self-examination of actions in the livestock unit, e.g. 'If I do this, will the animal suffer?' • Help and encourage less experienced stockpeople to develop their understanding and stock-handling skills

What owner/managers should do	• Discard belief (if it exists) that anyone can adequately care for animals as long as they are strong and fit
	• Establish and maintain contact, at both individual business and industry levels, with schools and colleges with the objective of influencing the curriculum content and careers advice on offer to young people
	• Develop a business strategy for attracting suitable applicants and selecting the best person for a stockman's post.
	• Develop a business strategy for retaining all competent stockpeople. Accept that a high rate of staff turnover is a sign of business weakness and a potential cause of inconsistent and poor animal care
	• Adopt a training and certification culture for the business and the industry. Offer appropriate initial training and continuing vocational training in animal husbandry knowledge and animal handling skills
	• Provide in the workplace clear-cut guidelines as to stockmanship practices which management does not condone on grounds of :
	   (a) cruelty to animals
	   (b) maltreatment of animals
	   (c) unpleasant treatment of animals
	• Ensure an adequate level of staffing in order to give a stockperson sufficient time to communicate with and get to know each animal
	• Include job satisfaction for the stockman as one of the business goals
	• Develop a business strategy for motivating each stockman as one of the business goals
	• Consider adopting a reward pay system that recognises excellence in stockmanship

- Act to enhance the image of animal stockmanship with the general public

What educators/trainers should do

- Promote the importance of high-quality stockmanship in relation to the attainment of high levels of animal welfare and productive efficiency
- Promote the value of continuous training as a means of attaining a high level of competence in stockmanship
- Continue to develop improved training packages in stockmanship
- Continue to improve on techniques and approaches in delivering these training packages

What researchers should do

Continue to:
- evaluate the role of stockmanship and its effects on various performance parameters
- develop improved methods of measuring the individual components of stockmanship
- investigate the role of individual components of stockmanship on the animal
- develop new and improved methods of training
- investigate how best the results of training can be properly evaluated

What the media should do

- Stop using disparaging remarks about the job of a stockperson
- Convey to good stockpeople the acclaim and status which is their due
- Improve the status of stockpeople in the eyes of the general public
- Highlight current shortcomings in relation to stockperson training
- Promote the benefits of training to the farmer, the stockperson and the industry as a whole
- Advertise the availability of training courses and packages and encourage active participation in these schemes by farm managers and stockpeople

What national bodies should do

- Take steps to discourage remarks which belittle the job of the stockperson and of stockmanship as a career
- Promote the image of stockmanship as a highly skilled and satisfying career
- Develop education-industry links at school level
- Canvass for development of recognised qualifications in stockmanship
- Recognise the importance of stockmanship when allocating research funds

Figure 11.2   A progressive route to excellence in stockmanship

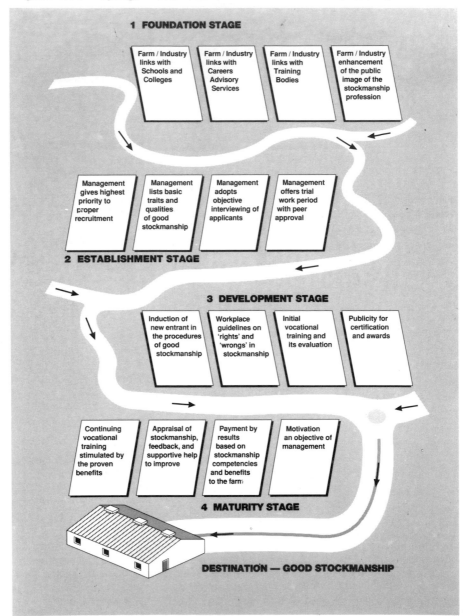

1 FOUNDATION STAGE

Farm / Industry links with Schools and Colleges

Farm / Industry links with Careers Advisory Services

Farm / Industry links with Training Bodies

Farm / Industry enhancement of the public image of the stockmanship profession

Management gives highest priority to proper recruitment

Management lists basic traits and qualities of good stockmanship

Management adopts objective interviewing of applicants

Management offers trial work period with peer approval

2 ESTABLISHMENT STAGE

3 DEVELOPMENT STAGE

Induction of new entrant in the procedures of good stockmanship

Workplace guidelines on 'rights' and 'wrongs' in stockmanship

Initial vocational training and its evaluation

Publicity for certification and awards

Continuing vocational training stimulated by the proven benefits

Appraisal of stockmanship, feedback, and supportive help to improve

Payment by results based on stockmanship competencies and benefits to the farm

Motivation an objective of management

4 MATURITY STAGE

DESTINATION — GOOD STOCKMANSHIP

definition is contained in Figure 11.3 (page 172). It is not a great deal different from the one outlined in Chapter 3 (pages 15–16). However, we have made minor additions and against each component we have summarised how improvements can be made.

One of the basic objectives of this book was to attain a higher status for stockpeople than they enjoy at present.

When stockpeople progressively develop their knowledge and skills and are adequately rewarded and motivated, they will not be the only beneficiaries of such a process, because, in turn, both the wellbeing of the pigs in their care and the interests of the business for which they work will be greatly enhanced in the process.

'Well, Mildred, can we now look forward with confidence to improved stockmanship?'

Figure 11.3 An improved definition of stockmanship

Component	*Main methods of ensuring high quality*
1. A sound basic knowledge of the animals and their requirements	Evaluation Training Experience
2. A basic attachment for and patience with stock 3. The ability and willingness to communicate and develop a good relationship with the stock (empathy) 4. Ability to recognise all individual animals and to remember their particular eccentricities	Select for stockperson posts those who: • have had experience with animals including pets • like animals • respect animals • have a strong desire to work with animals • demonstrate the characteristics of coolness, patience and tolerance towards animals and people Provide adequate time for stockpeople to attend to animal care
5. Keen sensitivity for recognising the slightest departure from normal behaviour of individual animals (perceptual skills)	As for 2, 3 and 4 above plus: ensuring that applicants have a high level of perceptual skills, i.e. good eyesight, hearing and sensitivity to the needs of individual animals and particularly to those in greatest need of care within a group
6. An ability to organise the working time well 7. Having a keen appreciation of priorities with a ready willingness to be sidetracked from routine duties when pressing needs arise to attend to individual animals	• Good sense of priorities developed through continuous training • Capable of adopting/evolving efficient working methods • Good judgement in allocating the appropriate amount of time to the range of jobs to be done • Keen ability, where necessary, to delegate some tasks to assistants
8. Work experience	A reasonable period of experience working with livestock combined with continuous education and training
9. Dedication to the job	• Providing motivation and ensuring good job satisfaction so as to encourage loyalty and therefore retention in employment • Interest in the animals • Persevering attitude • Determination to ensure that the job will be well done

ADDITIONAL READING

Additional Reading

Agricultural Training Board (1983) *Motivating staff*. ATB. West Wickham, Kent.

Anderson Y (1987) *A pilot study of the characteristics, attitudes and aspirations of stockmen*. Honours Degree Dissertation. Department of Agriculture, University of Aberdeen, Aberdeen, Scotland.

Argyle M (1978) *The psychology of interpersonal behaviour*. Penguin Books. Harmondsworth, Middlesex, England.

Armstrong J and Lloyd D H (1972) *Staff management audit*. Department of Agriculture Study No. 12. University of Reading. Reading, Berkshire.

Ballard D (1988) Stockmanship in pig production—Stockman recruitment, training and progression—The Pig Improvement Company Approach, January 1988. PIC. Fyfield Wick, Abingdon, Oxford, England.

Bennett, R (1989) *Managing people*. Richard Clay. Great Britain.

Beveridge L M, English P R, Burgess G, MacPherson O, Roden J A and Dunne J H (1992) Evaluation of the usefulness of and demand for training courses for pig stockpeople. *Proc. 12th IPVS Congress*. The Hague, Netherlands. August 1992. Paper 696.

Beynon N M (1988) Stockmanship in pig production—Meeting the industry's needs—Agricultural college courses. NAC Conference Paper, January 1988. National Agricultural Centre, Stoneleigh, Warwickshire, England.

Beynon N M (1990) *Pigs—a guide to management*. Crowwood Press. Marlborough, England.

Beynon N M (1991) Analysis of stockmanship. *Pig Veterinary Journal* 26, 67–77. Pig Veterinary Society, UK.

Brambell F W R (1965) *Report of the Technical Committee to enquire into the welfare of animals kept under intensive livestock husbandry systems*. Cmnd 2836. HMSO. London.

British Society of Animal Production (BSAP) (1980)   Summary of discussions and recommendations of the Pig Welfare Consultative Panel set up by the British Society of Animal Production, PO Box 3, Roslin, Midlothian, Scotland.

Central Statistical Office (1991)   *Annual Abstract of Statistics.* HMSO. London.

Cleary G V (1990)   Personnel management and staff training. In *Personnel Management and Record Keeping* 284–291. Pig production in Australia (eds. Gardner J A A, Dunkin A C and Lloyd L C). Australian Pig Research Council, Canberra. Butterworths.

Corson S A and Corson E O (1980)   *Ethology and non-verbal communication in mental health.* (Pergamon Press, Oxford)

Curnow B (1989)   Recruit, retrain, retain: Personnel management and the 3 Rs. In *Personnel Management*, November 1989, 40–47.

Curtis S E (1980)   Animal welfare concerns in modern pork production. An animal scientist's analysis. Paper presented to the Animal Welfare Committee of the US Animal Health Association at Louisville, Ky. National Pork Producers Council. Des Moines, Iowa.

Dryden A L and Seabrook M F (1986)   An investigation into some components of the behaviour of the pig stockman and their influence on pig behaviour and performance. *Journal of Agricultural Manpower Society* 1 (12) 44–52.

English P R and Macdonald D C (1986)   Animal behaviour and welfare. In *Bioindustrial ecosystems* (eds. Cole D J A and Brander G C). Elsevier Science Publishers, B. V. Amsterdam 89–105.

English P R, Fowler V R, Baxter S H and Smith W J (1988)   *The growing and finishing pig: improving efficiency.* Farming Press. Ipswich, Suffolk, England.

English P R (1991)   Stockmanship, empathy and pig behaviour. *Pig Veterinary Journal* 26, 56–66. Pig Veterinary Society, UK.

English P R, Burgess G, Bell A, MacPherson O, Beveridge L M, Roden J A and Dunne J H (1992)   Evaluation of the association between experience of and attachment to pets and empathy in pig stockmanship. *Proc. 12th IPVS Congress.* The Hague, Netherlands. August 1992. Paper 697.

English P R (1991)   The stockman's role in maximising sow welfare and efficiency. In *Proceedings Pig Improvement Company Conference* on 'The challenge of change—sow housing alternatives', October 1991. Pig Improvement Company. Fyfield Wick, Abingdon, Oxford, England.

English P R, Smith W J, Edwards S A and Dunne J H   *The sow: improving welfare, health and efficiency.* Farming Press. Ipswich, Suffolk, England. (In preparation.)

English P R and Edwards S A   Animal welfare. Chapter 72 in *Diseases of swine* (ed. Leman A *et al.*). Iowa State University Press. (In press.)

Freemantle D (1985)   *Superboss: the A to Z of managing people successfully.* Gower Publishing Company. London.

Gonyou H W, Hemsworth P H and Barnett J L (1986).   Effects of frequent interactions with humans on growing pigs. *Appl. Anim. Behav. Sci.* 16: 269-278.

Goodman S (1990)   Effective staff management. *Proceedings of Conference and Trade Fair 'Pigs North East'*, York Racecourse. (MLC/ADAS/NFU/ Farmers Guardian) March 1990, 12 pages. MLC, PO Box 44, Milton Keynes, England.

Grommers F J (1987)   Stockmanship—what is it? In *The role of the stockman in livestock productivity and management* (ed. Seabrook M F), Commission of the European Communities (EUR 10982). Luxembourg.

Hemsworth P H, Brand A and Willems P J (1981)   The behavioural response in sows to the presence of human beings and productivity. In *Livestock Production Science* 8: 67–74.

Hemsworth P H, Barnett J L and Hansen C (1981b)   The influences of handling by humans on the behaviour, growth and corticosteroids in the juvenile female pig. *Horm. Behav.* 15: 396–403.

Hemsworth P H, Gonyou H W and Dziuk (1985)   Human communication with pigs: The behavioural response of pigs to specific human signals. *Appl. Anim. Behav. Sci.* 15: 45–54.

Hemsworth P H, Barnett J L, Hansen C and Gonyou H W (1985)   The influence of early contact with humans on subsequent behavioural response of pigs to humans. *Appl. Anim. Behav. Sci.* 15: 55–63.

Hemsworth P H, Barnett J L and Hansen C (1986)   The influence of handling by humans on the behaviour, reproduction and corticosteroids of male and female pigs. *Appl. Anim. Behav. Sci.* 15: 303–314.

Hemsworth P H, Barnett J L and Hansen C (1987)   The influence of inconsistent handling by humans on the behaviour, growth and corticosteroids of young pigs. *Appl. Anim. Behav. Sci.* 17: 245–252.

Hemsworth P H and Barnett J L (1987)   The human–animal relationship and its importance in pig production. *CAB International Pig News and Information* 8: 133–136.

177

Hemsworth P H, Coleman G J and Barnett J L (1991) Reproductive performance of pigs and the influence of human–animal interactions. *CAB International Pig News and Information* 12: 563–566.

Herzberg F, Mausner B and Snyderman B (1959) *The motivation to work.* Wiley. London.

Herzberg F (1968) One more time: how do you motivate employees? In *Harvard Business Review*, Jan–Feb 1968.

Howard W H, McEwan K A, Brinkman G L and Christensen J (1990) *Human resource management on the farm: attracting, keeping and motivating labour on Ontario swine farms.* Department of Agricultural Economics and Business, University of Guelph. Guelph, Ontario, Canada. (Paper in Press.) (Summary of Masters Thesis of K A MacEwan.)

Kahan and Quinn (1969) In Warr P B, Cook J and Wall T D (1979). Scales for the measurement of some work attitudes and aspects of wellbeing. *Journ. of Occup. Psychol.* 52: 129–148.

Kiley-Worthington M (1990) *Animals in circuses and zoos.* Chiron's World. Little Eco-Farms Publishing. Basildon, Essex.

Levinson B M (1978) Pets and personality development. *Psychology Reports* 42: 1031–1038.

Lloyd D H and Armstrong J (1973) Monitoring human relations. In *Proceedings 9th Conference Farm Management Association.* National Agricultural Centre, Kenilworth, Warwickshire, England 1–10.

Lloyd D H (1975) Effective staff management. In *Economic factors affecting egg production* (eds. Freeman B M and Boorman K N). British Poultry Science, Edinburgh. 221–251.

Maslow A M (1954) *Motivation and Personality.* Harper & Brothers. New York.

Metz J H M (1987) The response of farm animals to humans—examples of experimental research. In *The role of the stockman in livestock productivity and management* (ed. Seabrook M F), Commission of the European Communities (EUR 10982). Luxembourg, 23–39.

Ministry of Agriculture, Fisheries and Food (Annual) *Agriculture in the United Kingdom.* HMSO. London.

Ministry of Agriculture, Fisheries and Food (1983) *British codes of recommendations for the welfare of farm livestock.* HMSO. London.

Mueller A G, Hollis G, Johnson L and Waldo M (1983) Employer–employee relationships on hog farms. In *Pork Industry Handbook* (PIH88). Cooperative Extension Service. University of Illinois, Urbana–Champaign, USA.

Mugford R A and McComisky J G (1974) Pet animals and society. In *Psychology Reports* 42: 1031–1038.

Muirhead M (1983) The veterinary surgeon's role as an adviser in pig production. *International Swine Update*. SQUIBB. March 1983.

Regnar Rylander (1986) Lung diseases caused by organic dust in the farm environment. *American Journal of Industrial Medicine* 10: 221–227.

Seabrook M F (1974) *A study of some elements of the cowman's skills as influencing the milk yield of dairy cows*. PhD Thesis, University of Reading, England.

Seabrook M F (1982) Functions of man in livestock systems. *Agric. Manpower* No. 4, 13–17.

Seabrook M F (1983) Human implications of modern livestock production systems. In *Stockmanship on the farm*. UFAW, Royal Veterinary College, Hertfordshire, England. 7–18.

Seabrook M F (1984) The psychological interaction between the stockman and his animals and its influence on performance of pigs and dairy cows. In *Veterinary Record* 115: 84–87.

Seabrook M F (1987) Research epistemologies—an holistic approach to investigating the role of the stockman. In *The role of the stockman in livestock productivity and management* (ed. Seabrook M F), Commission of the European Communities (EUR 10982). Luxembourg.

Seabrook M F (1988) The behaviour of the pig stockman and its influence on pig performance and behaviour—a review. *CAB International Pig News and Information* 9: 403–406.

Seabrook M F (1990) Reactions of dairy cows and pigs to humans. In *Social stress in domestic animals* (eds. Zayan R and Dantzer R), Kluwer Academic Publishers. London. 110–120.

Seabrook M F and Darroch A R (1990) Objective measurements of the suitability of individuals for livestock work and the implications for their training. 41st meeting European Association of Animal Production.

Segundo, Ricardo Cochran (1989) *A study of stockpeople and managers in the pig industry with special emphasis on the factors affecting their job satisfaction*. MSc Thesis. University of Aberdeen. Aberdeen, Scotland.

Segundo R C, English P R, Burgess G, Edwards S A, MacPherson O, Russell P A, Shepherd J W and Dunne J (1990) A study of factors which influence the job satisfaction of stockpeople on commercial pig farms. *Anim. Prod.* 50: 572–573.

Selye H (1976) *Stress in health and disease.* Butterworths. London.

Universities Federation of Animal Welfare (1982) Stockmanship on the farm. *Proceedings of a workshop. The Universities Federation of Animal Welfare.* Royal Veterinary College. Hertfordshire, England, Sept. 1982.

Vroom, Victor H (1964) *Work and motivation.* Wiley. London.

Vroom, Victor H (1989) (ed.) *Management and motivation.* Wiley. London.

Watson R D (1987) Pig housing and human health. In *Pig Housing and the Environment.* British Society of Animal Production. Occasional Publication No. 11, 41–51.

Wright D (1985) *Training investigation: pig production in Scotland.* Agricultural Training Board. West Wickham, Kent.

INDEX

Index

Colour plates, which are referred to by number, are between pages 114 and 115

FARMING PRESS BOOKS

Below is a sample of the wide range of agricultural and veterinary books published by Farming Press. For more information or for a free illustrated book list please contact:

Farming Press Books, Wharfedale Road
Ipswich IP1 4LG, United Kingdom
Telephone (0473) 241122   Fax (0473) 240501

The Growing and Finishing Pig   ENGLISH, FOWLER, BAXTER & SMITH

Explores in detail the many interlinked factors controlling the efficiency of pig growth from weaning to slaughter.

The Pigman's Handbook   GERRY BRENT

Describes in detail the best routines for the day-to-day running of a pig farm.

Outdoor Pig Production   KEITH THORNTON

How to plan, set up and run a unit.

Pig Ailments   EDDIE STRAITON

The visual signs of ailments with clear details of treatment and prevention.

Pig Diseases   DAVID TAYLOR

A detailed technical reference book about pig diseases for the veterinary surgeon and pig unit manager.

The Herdsman's Book   MALCOLM STANSFIELD

The stockperson's guide to the dairy enterprise.

A Veterinary Book for Dairy Farmers   ROGER BLOWEY

Deals with the full range of cattle and calf ailments, with the emphasis on preventive medicine.

A Veterinary Book for Sheep Farmers   DAVID C. HENDERSON

A wide-ranging, detailed guide to the prevention of sheep ailments, increased lamb output and the diagnosis and treatment of disease.

Farming Press Books is part of the Morgan-Grampian Farming Press Group which publishes a range of farming magazines: *Arable Farming, Dairy Farmer, Farming News, Pig Farming, What's New in Farming*. For a specimen copy of any of these please contact the address above.